Careers
in
Agriculture
and
Agricultural
Sciences

Careers in Agriculture and Agricultural Sciences

Susan Black

Kogan Page

First published in 1985 by Kogan Page Limited
120 Pentonville Road, London N1 9JN

Copyright © Kogan Page 1985

British Library Cataloguing in Publication Data

Black, Susan J
  Careers in agriculture and agricultural sciences
  I. Title
  630'.2'0341   S494.5.A4
  ISBN 0 85038 952 6 Pb

Printed and bound in Great Britain by
The Camelot Press Ltd, Southampton

Contents

Introduction   7

Part 1

1. **Agriculture**   11
   The Farming Regions 11; The Skilled Farm
   Worker 12; The Unskilled Farm Worker 13;
   The Farm Manager 17; The Self-employed
   Farmer 17; The Farm Secretary 18; Finding
   Work 19

2. **Crop and Food Science**   21
   Crop Science 21; Food Science and Biochemistry
   (with nutrition) 24; Vacancy Information 26

3. **Animal Science**   27
   Agricultural Zoology 27

4. **Care of the Land**   31
   Soil Science 31; Environmental Science 34;
   Agricultural Surveying 35

5. **Agricultural Economics and Marketing**   37
   Agricultural Economics 37; Agricultural
   Marketing 39; Vacancy Information 41

6. **Agricultural Engineering**   42
   The Operator/Mechanic 43; The Craftsman/
   Technician 44; The Professional Engineer 44;
   Future Prospects 45; Vacancy Information 47

7. **Other Careers Available**   48
   Fish Farming 48; Forestry 49; Overseas Work 50

Part 2

8. Courses and Qualifications   55
University First Degrees 55; CNAA Degrees 59;
CNAA Award 60; Higher Degrees and University
Diplomas 60

9. Diplomas and Certificates in England and Wales   66
Business and Technician Council Awards 67;
Higher National Diplomas 67; Ordinary
National Diplomas 68; Ordinary National
Certificates 70; Advanced Certificates 71;
Welsh Joint Education Committee Awards 73;
West Midlands Advisory Council Awards 74;
Awards from Other Bodies 76; Regional
Awards 77; College Awards 78; College Based
Awards 79; College Based Courses 80

10. Diplomas and Certificates in Scotland and Northern Ireland   82
Scottish Technical Education Council Awards 82;
National Certificates — Scotland 85; College
Based Awards — Scotland 85; Awards from
Other Bodies 86; National Diplomas — Northern
Ireland 86; Higher National Diplomas — Northern
Ireland 86; National Certificates — Northern
Ireland 86; College Awards — Northern Ireland 87

11. Useful Addresses   88
Universities 88; Polytechnics and Educational
Institutions Offering Degrees 89; Further
Education 90; Organisations Involved in
Agriculture 98

Introduction

Since the Second World War, agriculture, this country's oldest industry, has seen remarkable changes. The introduction of intensive farming techniques, mechanisation and automation has led to a more efficient, productive business and while some jobs have been lost at the manual level, others have been created elsewhere.

Advances in technology have led to the development of courses in the agricultural sciences. These produce scientists capable of carrying out research into improving food and raw materials production both in Britain and overseas and also of advising farmers and training them in the use of fertilisers, chemicals and suchlike.

However, practical farming is still very much a way of life and no fixed career structure exists in many areas, although the emphasis is now gradually being placed more and more on studying for academic qualifications and the development of technicians' posts in the agricultural sciences. Entry to many positions in agriculture is often dependent on a 'good secondary education'. Where this phrase occurs in the book it can generally be taken to mean education to O level or grade 1 CSE standard, preferably in five or more subjects to include mathematics and relevant sciences.

There is still more opportunity in agriculture for men than women, although this is changing in some fields, and those determined enough will succeed in their chosen areas.

This book looks at all the main areas of agriculture and its sciences and tries to give an idea of how a career should be pursued, together with the prospect of work after gaining practical and academic qualifications. In Part 1 of the book, Chapter 1 deals with career opportunities in general agriculture, Chapters 2, 3 and 4 consider different aspects of the agricultural sciences, and Chapter 5 looks at agricultural economics and marketing. Chapter 6 describes opportunities in the increasingly specialised field of agricultural engineering and this part of the book concludes with a brief survey of other jobs related to agriculture. Part 2 gives fuller details of courses and qualifications mentioned in Part 1 and also includes many addresses of organisations and societies that can provide useful information.

Although horticulture is a discipline related to agriculture, it is only briefly mentioned in the book and those interested in this area may find *Careers in Floristry and Retail Gardening*, published by Kogan Page, useful.

Part 1

Chapter 1
Agriculture

Agriculture uses up 67 per cent of the total land area in Britain for cultivation or grassland and grazing. Trends within the industry have led to fewer and larger farms, a reduced labour force that concentrates more on skilled workers, and the increasing use and application of science to improve production.

Farming still remains strongly traditional with close family links. In the past the vast majority of farmers and farm workers never went to agricultural college to gain specialist qualifications, but left school as soon as was legally allowed and returned to work on family farms. However, the younger generation of farmers are now becoming more qualified academically than their fathers and grandfathers.

Six different types of farming are recognised: arable or cropping, dairying, horticulture, livestock, pigs and poultry, and mixed farming.

The Farming Regions

The south-east of Britain is given over to mainly mixed farming, combining arable cultivation and livestock production, although separate farms for dairying and stock rearing, arable cultivation and sheep and pigs do exist. The south-west is still the stronghold of the small farmer and the area depends on sheep, beef and dairy production, with Wiltshire concentrating on intensive pig rearing. The Midlands relies on dairy, beef and sheep farming, and northern England is mostly given over to livestock as a lot of the land, being hill country, is unsuitable for heavy cultivation.

East Anglia and Lincolnshire are responsible for producing one-third of the nation's wheat and potatoes, two-thirds of the sugar beet, half the feed bean and vining pea crops, much of the oil seed rape and 80 per cent of the dried pea crop. It has become intensively cultivated and farms have become highly mechanised, leading to a steady decline in the number of farm workers employed in that area. Glasshouse crops and fruit are grown extensively as the land is of the highest quality. However, it is also an important pig producing area.

Wales concentrates on forestry and sheep and cattle production.

Scotland, by virtue of its climate, is an important producer of high quality seed crops as many diseases prevalent in England and Wales cannot survive the lower temperatures. Sheep and beef production is extensive, taking place on the mainland and islands. Scotland's best known crop, though, is timber.

The Unskilled Farm Worker

The basic function of the unskilled worker is to carry out all the manual work on the farm as well as operate simple machinery and equipment. On arable farms this would mean helping with ploughing, sowing and the management and harvesting of crops; on livestock farms it could involve the feeding and cleaning of animals; in dairying one would also assist at milking. Prospects for a full-time career are limited and those permanently employed may gain promotion to foreman only after long experience. Although older members of the farming community may have received promotion like this, it is not the best route to follow today.

The casual worker moves from farm to farm offering help at particularly busy times of the year. He would assist at harvesting on arable farms, moving on to fruit picking in the early autumn then back to field work at planting.

Agriculture

The Skilled Farm Worker

There is a recognised ladder of promotion here running up from skilled worker through foreman to farm manager. Career possibilities are enhanced by a combination of experience and study and an increasing number of employees take proficiency tests or certificates and diplomas in one of the six recognised specialist farming areas.

Arable Farming or Cropping

Ploughing, sowing, cultivating and harvesting are all the responsibility of the skilled worker, who can expect steady progress leading to management by building up experience and academic qualifications. The farming processes have gradually become more mechanised and advanced technology is used in crop management, so the arable worker will be proficient in the use of tractors, combines, grain dryers, potato harvesters and other specialised equipment. Courses can be taken on a day-release or full-time basis leading to national certificates in farm and grassland management, arable farming and other areas. National proficiency tests are also given in such areas as tractor driving.

Dairy Farming

The average herd size today is 40 cows and the dairyman is a vital factor in the smooth operation of the herd. Average milk yields are increased by controlled breeding, hence the wide use of artificial insemination, but just as importantly they depend on care and feeding which needs to be balanced and regulated. The dairyman/cowman (these are interchangeable titles) needs to know what is best for the herd, to recognise health problems, guard against infected milk and maintain production.

Cows have to be milked twice daily, so dairymen tend to work higher average hours than other skilled farm workers, including of course at weekends.

Most large dairy farms are highly mechanised and have sophisticated automated milking parlours. Cowmen have to be able to manage the equipment and keep it in working order, both from a mechanical and health point of view. They are also responsible for arranging matings or artificial inseminations, acting as midwives during calving, rearing calves, nursing sick animals, following vets' instructions and keeping accurate records of production and feed requirements for each cow.

Education and training in dairying is closely controlled by the Royal Association of British Dairy Farmers, who produce the syllabus and exams for the national dairying certificate. Colleges also offer TEC awards at various levels.

Livestock

Beef. The stockman's duties are concerned with the daily feeding, cleaning and general care of animals that are increasingly housed in intensive units. In breeding units mature cows must be cared for, breeding programmes organised, calving supervised and calves weaned. There must be careful monitoring of the herd to ensure production of prime cattle.

A national proficiency test in beef cattle stock tasks is a useful asset, as well as a college diploma.

Sheep. Sheep are either farmed on specialist upland farms or in small intensive units on lowland farms as a minor part of the output.

The shepherd on the upland farm needs to prepare for an active but very hardy existence, with no permanent indoor shelter, as sheep live in the open all year round, roaming extensively over wide areas of rough hillside or moving from pasture to pasture in the lowlands. With the help of one or two trained dogs he will have total responsibility for the flock. He must be expert in breeding the right sheep for the environment, knowledgeable in caring for ewes, and lambs at breeding time, quick at spotting injuries or disease and pests and know how to dip and shear sheep and maintain the flock.

The best academic base for the shepherd is a national or college certificate as well as proficiency tests.

Agriculture

Pig Production

Pig farming is a highly specialised intensive business that can be very profitable, but production has to satisfy the exacting requirements of the market for high quality and lean bacon and pork. The national certificate is again the best route to progress in this area.

Pigs are normally kept in special units either on large mixed or stock farms or holdings, and skilled pigmen need to be expert in their care and nutrition, health and breeding as well as being able to keep accurate records of weight and food intake.

Poultry

In the past 20 years the poultry industry has developed from a sideline into a major concern — a rise due to technological advances in intensive battery farming. An employee here will care for 60,000 or more birds with the help of modern aids, looking after broiler rearing and laying birds who produce eggs for sale and for hatcheries.

Poultry men and women work indoors in controlled conditions and have irregular hours. Although the job is often lonely, the trained poultry worker has great responsibility: the quick recognition of signs of distress, poor hygiene and feeding problems, as well as the maintenance of accurate production records, are essential to economic efficiency.

One can enter the business with a good secondary education and expect to advance from trainee stockman to senior stockman and on to unit manager and then production manager. Many poultry companies will organise training through day release at local colleges, and help with national proficiency schemes.

Farm Machinery

Machinery now plays an increasingly important part in any career in agriculture. The Agricultural Training Board recognises this and runs courses in machine handling and tractor driving, as well as teaching the use of highly specialised equipment, eg pea viners and beet harvesters.

Mechanics are trained as engineering apprentices, and further information on this field can be found in Chapter 6, which deals with agricultural engineering.

Careers in Agriculture

Contract Work

Agricultural contractors employ experienced workers and managers to provide specific help to farmers at certain times of the year. About 60 per cent of all farms use contractors from time to time. Although contract work is not undertaken in this country to the same extent as in Holland and the USA, it provides skilled workers with the opportunity to work in a variety of different locations from day to day.

This is not so much a career as a way of gaining useful, all-round experience not found on the individual farm, and those seeking prospects of promotion are best advised to secure permanent positions.

Case Study

John is 26 and after leaving agricultural college with a certificate in general agriculture, as well as having a farming background, he joined a large company involved with sugar beet production and processing. After beginning as a trainee he is now a fieldsman, with many responsibilities.

> On joining the company, I was put through an intensive 18-month training programme which involved working on the cultivation and harvesting of beet in the field and learning about current technical advances and research.
>
> I was automatically promoted to fieldsman on completing the course. I now advise the farmer on how to grow his beet profitably and efficiently and to keep up the highest standards of root quality. During harvesting I will help to organise contract labour and co-ordinate delivery to the factory.
>
> My year is quite varied, beginning in October when planting and the first stages of plant development need to be carefully monitored. I have to be aware of new seed varieties and developments in pest and weed control. During its early stages, beet is particularly susceptible to poor growing conditions and pests and diseases, so I have to be on hand in the field to prevent crop failure. When the crop is established, I begin to plan the harvesting and delivery and this brings me full cycle.
>
> Career prospects are good with gradual progression to management posts, and I get great satisfaction from my direct and frequent involvement with the practical work in the field.

The Farm Manager

The main function of the farm manager is to carry out the owner's policy for the farm, supervise staff, organise field and other work, market the produce successfully and keep accurate records. Farm managers are either employed by the private owner or by co-operatives and corporations that have extensive farming interests. Career prospects are good, especially in large corporations where one can progress to control a number of farms.

One way of becoming a farm manager is to work up through the ranks as a skilled worker, but the more recognised route is now to have an academic training in the form of a degree or a diploma from agricultural college.

Academic Training

It is possible to train either full time at agricultural college or via a three-year apprenticeship scheme with day- or block-release for study. One normally needs to have a good secondary education for acceptance.

For school-leavers with two or more A levels from either the sciences or mathematical subjects there is an option to study for a degree in agriculture or, for those wishing to enter practical farming, to take a national certificate.

At all levels practical experience is invaluable and it is often quite difficult for someone from a non-farming background to gain a post even after academic training. Individual colleges and universities should be consulted on their requirements.

The Self-employed Farmer

Unless one inherits a farm, the main problem facing those wishing to become farmers in their own right is finding sufficient capital. Land prices keep on rising and fewer farms or smallholdings are becoming available, with large companies able to outbid offers. Becoming a tenant farmer used to be the traditional way of entering farming, but this is no longer possible to the same extent as less property is being rented out.

To buy and stock even a small farm requires considerable capital, plus a thorough training in agriculture and business management. Some can become partners or farm directors, but the best course of advice for the aspiring farmer with little capital is to become a farm manager.

The Farm Secretary

The farmer has to keep extensive records and deal with correspondence, sales representatives etc, so the farm secretary is an important member of the farming team in keeping the administration in order. He or she is made responsible for general farm business, accounts, VAT returns, wages, grant applications and subsidies, and records.

Most large farms employ full-time secretarial help but it is common to find a group of small farms employing the same secretary, who will travel from farm to farm, depending on individual needs for his/her skills. Many such secretaries work for agencies.

There are special courses available for secretaries. The new BTEC national award in business studies is now replacing the old ordinary national diploma (OND) and a special national certificate is also available.

Case Study

Elaine is 24 and works as a farm secretary on a large mixed farm owned by a group of companies. Her qualifications are a livestock farming background and a college diploma as an agricultural secretary. In addition, she has been taking A level accounting at night school plus a course in computing. All these are essential to the job. She is accountable to the farm director.

> My duties as farm secretary involve responsibility for keeping farm records, VAT returns, all cheques written and the petty cash, wages and National Insurance for the men, and arranging appointments for sales reps with the farm manager. Recently a computer system has been installed in my office which I take responsibility for, deciding what information

is put on to disc storage. As a sideline I also help with calf rearing.

I normally work alone in the office, from 8 am to 4.30 pm, but when involved with the calves I work from 7 am to 6 pm, and a few hours at the weekends, feeding them twice a day. I find that my time in the office is divided fairly evenly between the responsibilities of data processing, conferring and advising, and planning.

My satisfaction comes from being able to use my initiative and be actively involved in the farm and, although promotion probably does not exist as such, my responsibilities will increase.

Finding Work

The best source of information for vacancies in general agriculture is the *Farmers Weekly*, which should be available at town libraries and is on sale in most newsagents. Casual labourers, unskilled and skilled workers and farm secretaries may also find posts through local newspapers, the national press and Jobcentres. University careers services and agricultural colleges provide lists of appointments available to those suitably qualified.

Youth Training Scheme

Of relevance to 16- and 17-year-olds who have left school and would like to pursue a career in agriculture, but lack the experience, is the scheme recently set up by the Manpower Services Commission.

The Cambridgeshire Farm College, Milton, Cambridge, has just finished the first year of a one-year course based on agriculture. Normally the college only trains people already employed in agriculture on a day-release basis for the City and Guilds of London Institute exams, but it is acting as the training agent for the MSC-funded Youth Training Scheme. Last year, 19 people joined the course which lasts 50 weeks and includes 13 weeks of theoretical training. Last summer, 15 of them had taken jobs in farming and two had gone on to further study. Five people were given the chance to study agricultural mechanics and all became fully employed as a result.

The course in agriculture includes training in the maintenance of water supplies, dry-stone walling, workshop practice, gate hanging and the care of hand tools, and is very thorough. Both males and females can attend.

Although this scheme only operates in Cambridge at the moment, those interested should contact the Manpower Services Commission to find out more details and see if any other colleges around the country will be holding similar courses.

Chapter 2
Crop and Food Science

The agricultural industry is constantly looking for new ways to improve yields and the quality of crops and crop products. Demand for research and development has in turn led to the teaching of specialist courses to cover the science behind crop and food improvement.

This is a rapidly expanding field, offering great career potential. Many skills are involved and most courses at all levels concentrate on giving students a substantial grounding in biological sciences, chemistry and mathematics.

Crop Science

This area includes plant science, agricultural botany and crop production. The crop scientist is primarily concerned with exploiting the genetic potential of a crop to produce varieties of higher yield expectation, with higher disease resistance, frost tolerance (in winter-sown crops) and other desirable factors. He or she will also apply the principles of management to the spreading of fertilisers, pesticides, fungicides etc, to the best advantage and in the most economical manner.

As with all the agricultural sciences, most teaching begins at degree level to produce qualified scientists. Some, however, are able to enter crop science by taking agriculture courses at college that specialise in arable farming and production, or join companies as trainee scientific staff and continue their education at local technical colleges taking agricultural diplomas or certificates.

Careers exist in a variety of fields and offer a wide range

of posts and responsibilities. The pattern of promotion described here also applies to posts in animal science and soil science, covered in later chapters.

The government provides funding for many research establishments and examples of those involved in crop research are the Plant Breeding Institute, Cambridge; the Glasshouse Crops Research Institute, Littlehampton; and the Grassland Research Institute, Maidenhead. The number of vacancies each year varies and can be heavily dependent on the size of grant available. In places such as these, one follows the career structure common to all scientific positions in the Civil Service. Those with BTEC higher diplomas/certificates or a good secondary education are normally taken on at scientific officer level. Promotion to the posts of higher scientific officer and senior scientific officer should follow with further training and experience, and comes almost always from inside. The same scheme for appointments exists within the National Institute of Agricultural Botany, which is primarily involved with the testing of new crop varieties. There are also advisory posts in the Ministry of Agriculture but open only to graduates, preferably with experience; promotion is again on civil service lines.

In the private sector, many large companies own their own farms and employ graduates or holders of appropriate certificates in research and development divisions. Some will run trainee schemes and employ people with a good secondary education. Here one would expect to begin as a field trials assistant and gain promotion to senior managerial positions, such as becoming head of a trials unit. Other companies may employ advisers on crop science, and from a graduate recruitment scheme one can expect to reach senior advisory positions. Overseas development organisations provide a popular outlet in this direction although many candidates have a higher degree to offer for such jobs.

Case Studies

Judith is 24 and has just been promoted to head of crop science in a large private company that has an extensive research division. Her basic qualification is a degree in plant science, which has been essential to the job. She is responsible for six other full-time staff and up to five students during harvesting in the summer months.

> On joining the company I was attached to the cereal agronomist as his assistant and spent six months under his direction, conducting field trials and evaluating yields of bread wheats and the disease resistance of new varieties bred by scientists in the firm.
>
> I then led similar projects, including ones evaluating the effect of different levels of fertilisers on crop varieties, in conjunction with other departments and local farmers. I completed my training under the direction of the agronomist and was given total management responsibility for the field trials sites. This gave me far more contact with other research organisations and commercial companies, in giving advice and undertaking contract trial work.
>
> I was recently promoted to the position of head of crop science. This has led to a change of emphasis in my work. I spend less time than before in the field and more time supervising field staff, promoting relations with local farmers and making information on our work readily available to farmers and other interested parties. I also tend to have more contact with the laboratories who process the material harvested in the field, and have to be aware of improvements that can be made in my department. I must now concentrate equally on all these areas of my work.
>
> I work fairly set hours in the winter months, 8.30 am to 5.30 pm, but in the summer months remain flexible to supervise trial harvesting and data recording.
>
> I have found my work in the industry challenging and rewarding so far and my prospects here seem good. I enjoy being able to work outside, the frustration being that the more promotion I receive the more time I need to spend in the office.

A 28-year-old, who is now a mill manager for an industrial group of companies with extensive grain interests, describes his present responsibilities and the training involved to get there.

> I joined the group as a production management trainee and

spent my first six months with a senior member of staff at a newly opened mill, learning the basics of the grain industry on the job.

After this initial period I was moved to another mill and given an opportunity to show my managerial potential by being responsible for organising day-shift rotas and supervising night work. At the time I thought this was too much to cope with too soon, but with invaluable help from my colleagues I succeeded. I also found I was gaining technical knowledge in this situation.

At this point in my career, the company decided to keep abreast of technological advances and install computer systems in a few mills to see if efficiency was improved. I became involved with the commissioning, installation and staffing of the system in my mill, and saw the project through to operation. I then spent a year helping with computer installation in other mills.

Some months ago, I was appointed mill manager. I now take on the total responsibility for the running and operating efficiency of my site. The bulk of my time is thus spent on production overseeing and development, personnel work and meetings, along with some research as well as attendance of training courses. My training here, and lecture courses at various institutes, has supplemented my degree in agricultural botany and my agricultural background, both of which have also been very useful. I now have the chance to further my career to higher managerial positions.

Food Science and Biochemistry (with nutrition)

This subject is taught as a separate discipline and concentrates more on the chemical, biochemical and nutritional aspects of food production. The food scientist is involved with the value and quality of consumable products together with the processes affecting them. He or she will be concerned with microbiology here, a science fundamental to so many production methods, for example yoghurt and cheese making and brewing, as well as the correct packaging, storage and testing to protect food from moulds and prevent poisoning.

Most careers are in research and development in private food companies or government departments. The brewing and dairy industries offer a variety of posts, and catering is also full of opportunities. Places in advisory work are available and here it is possible to become a public health officer, working in schools or companies monitoring the handling of produce.

In research, training opportunities exist at various levels. Holders of BTEC higher awards can find openings at assistant level and then progress to technical scientist or senior technical scientist. Graduates gain appointments as research scientists, moving on to senior or principal grades. There are university courses available in both agricultural food science and agricultural biochemistry with nutrition; food science courses at the BTEC higher levels are taught as parts of other main courses. To find out which ones are suitable, contact the Business and Technician Education Council at Central House, Upper Woburn Place, London WC1H 0HH, explaining what you hope to do after study.

Case Study

A 23-year-old describes his work in food production. He is employed in the dairy division of a company specialising in milling and baking, dairy products, meat processing and brewing. He is now a process manager.

> I first joined the company at one of their main dairies, as a technical assistant. Here I learnt about the structure of the dairy industry, the handling of milk and milk products, and their treatment, packaging and distribution. I was particularly involved with the production of butter and cheeses. At the same time, I took a great interest in new developments in dairy technology and this led to promotion to assistant dairy manager.
>
> I then became responsible for the installation of a new computerised dairy-processing machine, and for the update of our cartoning methods. Once installed, the machinery led to increased production and fewer errors.
>
> My rapid promotion to process manager came as a welcome surprise but was related to this success with the equipment. This has meant I am now in charge of the bottling and cartoning of milk, buttermilk and yoghurts. The variety of problems I can be faced with is enormous. It can range from investigating the supply of a dubious tanker of milk to a customer complaint of sour milk. Machine breakdown may mean reorganising staff for overtime work, calling in the engineers and switching production plans around.
>
> I have become increasingly involved in staff management

and form the link between senior management and the production workers, trying to implement their instructions to the best possible advantage for all. To be successful, I have to create a team of people who want to maintain the high quality of dairy products we make and keep the best standards of hygiene. This requires me to be flexible towards my staff and forward thinking. I hold once-monthly meetings where we discuss the business and new product ideas.

My future career will extend into higher management or to a senior technical advisory post, but at the moment I get enjoyment in seeing a job well done and having a happy work team. I now have an extremely varied range of tasks with most of my time being spent on production, maintenance and hygiene checks, but also some research, staff and management meetings and product testing and technology updates.

I entered the dairy with a degree in food science and would say I have found it very useful.

Vacancy Information

For those people qualifying as technical and scientific specialists in any of the agricultural sciences discussed in Chapters 2, 3 and 4, *Farmers Weekly* and the national press are valuable sources of vacancies. College and university career services should be consulted, as should faculty or departmental notice-boards, as some posts are not advertised nationally but may be drawn to the attention of academic staff who then pass information on to the students.

For those wishing to pursue careers in research, useful journals available in university and public libraries are *New Scientist*, *Nature* and the American-based *Science*.

Chapter 3
Animal Science

Many forms of livestock are now farmed intensively, with much attention given to animal hygiene, diet, feed-to-weight conversion efficiency, and the breeding of superior-quality animals. The EEC legislation in agriculture has had a sharp effect on the number of dairy herds in this country and the production of lamb for the home and foreign markets. The housewife has become more choosy in the type of meat she wants, as has the catering industry, demanding leaner carcasses.

Livestock farming has been keeping up with these changing demands in the market-place, by using modern technology to improve its product. Animal scientists are trained to give advice and help for the benefit of the farming community.

Agricultural Zoology

The agricultural zoologist is trained in all aspects of livestock production, from the maintenance of the animal to its preparation as a food product. The health and well-being of animals are the priorities in this field and government departments and private companies are chiefly concerned with developing drugs, animal feeds and chemicals for crop production. Courses are designed to give weight to the physiological and biochemical aspects of animals, and a general grounding in agriculture is also provided along with an emphasis on the value of proper feeding and health care.

Some training is provided at BTEC level, but direct entry to most scientific posts is through a degree. There are courses available in general agriculture, agricultural zoology, animal nutrition, physiology and the biochemistry of farm animals. At first degree level, students are unlikely to specialise in one particular form of livestock to a great extent, but will learn about all types. There are also many higher degrees available with a greater level of specialisation. Agricultural colleges also hold courses biased towards animal science and these can lead to posts as animal technicians and assistant research workers.

The major areas of employment are within government financed or privately run research and development departments, in advisory work, public health laboratories and meat inspection, as well as animal hygiene. For those specialising in nutrition, many companies offer posts as production managers in feedstuffs factories. Some can find careers in field trials work, but these are conducted on a smaller scale than crop trials because there are problems in animal uniformity (obtaining an average feed conversion rate when every animal is different) and with the collection and processing of data. Other areas include marketing and sales posts in feed companies or overseas advisory work.

Those with a good secondary education can either become research assistants and attend day-release courses at a technical college, or take up an assistant scientific officer position in other areas, or they can take a relevant course at agricultural college before joining at a slightly higher level. Graduates can expect to be appointed as scientific officers or the equivalent grade in other areas of animal science.

Although there are still opportunities for those without degrees or college certificates, such posts are becoming fewer, and most places prefer graduate entry especially for scientific work.

Case Studies

Anne is 23 and has worked in the research laboratory of a

large pharmaceutical company for two years as a research technician. Her main role is to look into the drug resistance of certain parasites affecting farm animals, as well as different projects run with other organisations.

> When I joined the company I was given appropriate training to enable me to hold a Home Office licence, giving me responsibility for all on-site experimental vaccinations and other animal work. In my present job I have a dual role, where I take charge of all the parasitology work carried out in the laboratory and also co-ordinate any field trials being carried out on three nearby farms. I work with two other qualified vets, three other technicians who hold degrees, one technician who holds a BTEC certificate and a general lab assistant.
>
> My practical time is divided equally between the lab and the field, with the additional administrative work of planning, data processing and writing being done in the lab. I also spend a fair amount of time conferring and travelling. In this line of work it is important to be able to mix well with people — for farm trials and consultations — to be confident with animals and be flexible, allowing for a complete change of plans in order to accommodate the unexpected. Disadvantages of the position include a lowish salary, and not being able to act without higher authority, but my career structure will eventually lead me on to project management and developmental control posts.
>
> I joined the company after reading agricultural zoology at university and gained experience working in veterinary laboratories during vacations. All this has been very useful to me in my work although I do not regard my degree as essential. The company has provided me with further training where needed and I have attended a number of short courses at veterinary schools covering large-animal vaccination, blood sampling, poultry diseases and animal management. My colleagues in the lab taught me techniques I did not acquire at university.

Peter is 25 and joined an agricultural firm that specialises in selling fertilisers, animal feeds, crop chemicals and pharmaceutical products to farmers, as well as buying grain and other goods in return. Peter began as a trainee sales representative and has since taken promotion to a specialist post, controlling animal-feed quality and distribution.

> On joining the supply firm in August, I was placed in a

regional sales office and helped with grain collection and sampling over the harvest period. I was then given a three-month intensive sales training course which covered all the products we made, sales techniques and sales administration. Most of the course took the form of lectures and discussion groups, but some time was spent visiting distribution and production centres that stored goods I was to sell. I also had responsibilities in the field and meeting farmers.

On completing my course, I was given an established sales area to control. I was introduced to existing customers by the representative I replaced (he had gained promotion to development officer). A client record system was passed over to me which enabled me to know the general needs of each farmer in my area and discuss appropriate products on future visits. The company provided a car for both work and private use, which was a big bonus, and gave me a telephone allowance for work conducted at home (clients having both my office and home numbers). The area I covered contained approximately 200 farmers and I had to arrange to visit each one every four to six weeks. I had to maintain if not improve sales targets in my area. The work at times became lonely and I often felt depressed as I had little time to socialise, but as farmers began to know me I was looked on as a source of advice and help with feed problems.

Eventually I followed up my interests in nutrition and was promoted to sales specialist in animal feeds in the area. My job now does not involve direct selling as such, but is more an advisory post, keeping the sales force fully informed on feed products and helping farmers with their livestock's feeding difficulties. Farm visits form about a third of the work, while the rest of my time is divided fairly evenly between research, conferences, meetings and arranging trial work.

While finding my time as a sales representative useful I am far happier in my current post. Career prospects are good with advancement coming to senior feed representative and other technical posts, such as field sales manager. My major frustration with the career structure is that the company insists on one joining the sales force before allowing specialisation in the scientific or technical aspects of animal feeds.

I came into the job with a degree in animal nutrition and feel this was essential in obtaining my present position; earlier the company gave me the full product and sales training that I lacked. I feel it would have been an advantage to have come from an agricultural background, as farmers remain suspicious of sharp sales reps who can repeat what they learnt during training but have little in-depth knowledge of farm problems and needs.

Chapter 4
Care of the Land

The soil is fundamental to both livestock and arable farming, providing the base for crops to grow in. As pasture and arable land is farmed more intensively, the old-style crop rotation with a four-year cycle of growing different crops for three years and leaving land fallow in the last, to rest the soil, is now uneconomical. However, constant land use quickly leads to a breakdown of soil structure and a large loss of minerals and other constituents. To help us understand the soil and its needs, soil science is now taught, looking at all aspects of the earth and the improvement of its fertility.

Interlinked with all the agricultural sciences is the study of the environment and ecology. The long-term effects of using high levels of artificial fertilisers, and the problems caused by removing hedgerows to increase acreage and ease mechanical handling are two of the many varied problems the environmentalist has to study. Just as important as improving our farming production is the maintenance of a natural balance in the countryside to the mutual benefit of all.

To complete the agricultural sciences a brief look is taken at agricultural surveying. This too is becoming a specialised discipline taught to a professional level.

Soil Science

This is taught at first and higher degree levels, training scientists in the study of the soil, and soil-plant and soil-machinery relationships. Courses include analysis of soil types, their suitability to different cropping and cultivation, the use and effects of fertilisers, as well as soil erosion.

The vast majority of soil scientists find careers in advisory work or research and development. The Ministry of Agriculture employs soil advisers in a number of places, and here graduates will begin as scientific officers and gain promotion to higher scientific officer and senior scientific officer levels. Research workers can find employment in publicly funded or private establishments. For example, Rothamstead, an experimental research station, provides soil physics laboratories and carries out soil chemistry tests and research. Some research workers can become involved in overseas work.

In the private sector, agricultural companies have vacancies in similar posts and career development can be expected along comparable lines to the public sector, with scientists moving up to senior advisory or managerial posts.

A few graduates go to work for land consultancy companies. These mainly advise on the reclamation and redevelopment of land disturbed by quarrying. Two types of quarry exist — hard rock that includes china clay pits in Cornwall, and limestone quarries — and sand and gravel pits. The government departments tend to look after restoration of coal pits and coal waste tips. At a junior level one can expect a heavy involvement in field work, which will lessen as more senior consultancy roles are taken on.

Case Study

Pat is 24 and works for a small land consultancy company in the south of England.

> I joined the company from university as a soils consultant following a temporary job with the Soil Survey of England and Wales on a contract soil survey in western England.
>
> My first few months at the company were spent under the supervision of senior staff, particularly the senior soils consultant. I accompanied colleagues on various site visits but was responsible from a very early stage for soil survey and land classification work and was expected to contribute usefully to existing reclamation projects.

During this initial period I was assigned my own clients and became responsible for all work associated with their sites. I was also expected to follow up any contacts that would lead to further work. I thus received very little formal training, mainly because in a small commercial firm there are not sufficient resources for a lengthy period of training. However, the whole firm works very much as a team and help from other colleagues is always readily available. I now work on my own projects and sites but I frequently consult with workmates, and several members of staff may be involved in the submission of a complete land restoration scheme. Although certain sites may be my personal responsibility, work through the year tends to be somewhat erratic and there are often times when colleagues help out and vice versa.

The company is involved in a wide variety of agricultural and environmental projects ranging from herbicide trials and crop litigation (law suits involving damage to crops) to problems of waste disposal and contaminated land. However, a major part of the work is in land restoration, particularly restoration to agriculture after mineral working. The company provides consultancy services at all stages from pre-development surveys through to restoration design and planning applications, and finally implementation and monitoring of a scheme.

There are eight full-time staff, of whom six are technical and two are secretarial/administrative. All the technical staff are honours graduates qualified in agriculture, agricultural engineering, ecology or soils. Three have a higher degree (PhD or MSc), though this is not essential and relevant work experience is considered more important for employment.

Most of my work is concerned with site visits, reports, planning applications and advisory letters. The rest of the time is spent travelling and in dealing with administration and background research. Career prospects within the company are limited by its small size but are compensated to some extent by the wide range of experience to be gained, as well as the chance to work outdoors, tackle a variety of projects and meet many interesting people. However, I can foresee a time when promotion would mean a change of job, and there are the additional frustrations of having to justify all actions taken on behalf of a client, along with providing a detailed breakdown of costs to a client when the bill for the completion of a project is sent out. Additionally, I am not always able to devote enough time to potentially interesting research.

My initial qualification in soil science has proved essential and previous experience with the Soil Survey, although brief, was invaluable. Nevertheless, I have had to acquire a knowledge of many associated disciplines and indeed other skills. A recent example of the latter was organising and running a two-day residential conference for 30 people.

Environmental Science

This is taught mainly at degree level, although some colleges may include a short ecology section in their teaching. The aim of specific training is to produce scientists who are aware of the problems facing our environment today, and who seek to implement a conservation policy that benefits people and nature alike. Courses include sections on pollution, the use of land for leisure and recreation, cultivation, public health, national heritage and community services and housing.

The types of career available vary considerably. As yet there are no definite career paths to follow although with most posts one can expect to begin as a conservation officer, and be personally involved in field work, and then receive promotion to posts of a more managerial kind. Here one would act more as co-ordinator and administrator, with a work-force to take care of. The most likely sources of employment lie in the following areas:

☐ Large industrial and agricultural companies who employ environmental advisers to give help to farmers and company employees on the control of pollution at work, from the manufacture of chemicals and the effect of waste on surrounding land, to monitoring the level of fertiliser components coming off the land and entering water systems.

☐ The Ministry of Agriculture takes on advisers who follow the career pattern of all Civil Service scientific staff.

☐ The Nature Conservancy Council employs advisers and information officers. The Council has just launched a new initiative on public awareness of the problems facing us, and uses information officers to inform the farmer and the man in the street what is being done, what needs to be done and how we can all contribute.

Care of the Land

☐ The Farming and Wildlife Advisory Group recruits conservation officers, who co-ordinate field projects and give advice.
☐ The Institute of Terrestrial Ecology and the National Trust have similar projects.

Educating the general public plays a large part in environmental work and some officers find a teaching certificate a useful asset. Degree courses are available in agricultural and environmental science, ecological science, environmental biology and medicinal, agricultural and environmental chemistry. There are also relevant higher degrees.

It should be remembered, though, that environmental studies continues to be a popular discipline, meaning strong competition for careers, especially at lower levels where employers can select from a number of well-qualified graduates. To be able to offer practical experience can be an advantage. Many of the organisations mentioned above take on volunteers at weekends and during vacations to help on a variety of conservation projects.

Agricultural Surveying

This is a branch of surveying mainly involved with the management of rural land and buildings, along with woodland, parkland and scrub or waste land. It also includes advisory work on the use and development of existing property and the acquisition of land for investment.

Traditionally, land agencies were responsible for managing landed estates but these have now expanded to include agricultural surveying, giving advice on specific problems affecting rural property. This in turn has to include tenant rights, rating and valuation of properties, negotiation of claims for compulsory acquisition, stock-taking valuations, and design and alteration of farm buildings.

A rigid professional training scheme is operated to produce qualified surveyors and technicians, and the career requires one to have certain qualities and attributes. Surveyors need to be numerate, possess or develop business acumen, and be accurate in their work, as the smallest mistake can be very costly. Although the surveyor's role is essentially one of giving advice, and therefore lacks direct involvement or creativity, he or she must have the ability to see the overall picture and understand the full implications of their work and advice. A wide background knowledge of many other disciplines is needed.

The work is mainly out on site, involving a great deal of travel and extensive consultation and co-operation with other professionals.

To become a qualified surveyor, one can take the old-established professional training route and join a company as a trainee, taking college training in a particular post leading to final exams. Alternatively, there is the opportunity to take a degree course and enter at a higher level on the training scale. The land economy tripos at Cambridge is still recognised as the best of degree courses in this area, but other universities that include agriculture in surveying courses are Aberdeen, Edinburgh, London (Wye College), Newcastle and Reading. The College of Estate Management has now been incorporated into Reading University.

Polytechnics and the CNAA are developing more courses, with estate management available at Bristol, Leicester, Liverpool, South Bank, Thames and Wales Polytechnics. Most of the training takes the form of full-time or sandwich courses. The Royal Agricultural College at Cirencester offers its own specialised course. The Royal Institution of Chartered Surveyors encourages the teaching of BTEC awards and those suitably qualified can go on to obtain professional status.

As this is still a developing area, for full details on technical and professional qualifications together with career prospects in agricultural surveying, one should contact the Royal Institution of Chartered Surveyors and the Business and Technician Education Council.

Chapter 5
Agricultural Economics and Marketing

Economics finds applications in all industries and successful farming relies increasingly on sound financial principles to keep it profitable.

Farm turnovers, from the smallest family concern to the large limited companies, are generally on a much larger scale than in other industries, as machinery, fertilisers, buildings and livestock require a huge capital outlay. As a consequence, competition in the market-place between the agricultural supply industries for the farmer's custom has led to recruitment of trained salesmen and marketers to win business. There is a recognised skill in selling products and to be successful depends on good market strategy. Agricultural marketing has thus been developed as an academic discipline, with students being taught the theory of economics and how this is linked to the practice of marketing. As vast amounts of money are involved, more and more companies are looking at the graduate field to fill vacancies.

Agricultural Economics

The main function of the agricultural economist is to ensure modern business methods and applied economics are put to the best use to aid the farmer.

Specialist teaching really begins at degree level. Two basic methods of teaching operate in the universities. Aberdeen, London (Wye College) and Reading put the main emphasis on agriculture and lecture within that department, whereas Edinburgh, Exeter, Glasgow, Manchester,

Newcastle, Nottingham and Wales (Aberystwyth and Bangor) base studies in the economics department, with agriculture taught as a subsidiary subject. Higher degrees are also offered at a number of universities.

The majority of graduates stay within agriculture and of those who do not take up academic posts or go on to higher degrees, most find careers in the Civil Service — in various governmental departments — in commerce, in advisory work in both the private and public sectors and in sales management. A few gain posts abroad either in sales or advisory roles. Very few opportunities occur for school-leavers but a few companies may take people offering good A level grades on to trainee schemes and give in-house training supplemented by a technical college course in basic economics. However, promotion may be slow and in favour of graduates.

Case Study

Paul is a 30-year-old branch trading manager for a private group of food companies in the United Kingdom and he works mainly by himself. His initial qualification was a degree in agricultural economics and as he did not have an agricultural background he found this very useful. Whatever his degree did not cover the company taught once he was working.

> When I first joined my company, I was given 12 months' comprehensive training in all aspects of the malting barley trade (which was my chosen specialist area). To supplement the company lecture courses, which were given by agronomists, economists, farmers and maltsters, I accompanied experienced traders for practical instruction in grain sampling, and all the processes involved in barley handling from the farm at harvest to delivery at the mill. Without this inside knowledge of the operations involved, I would have found it difficult to carry out my present job effectively.
>
> I then spent two years buying and selling grain and other products in a specified area of the country. At this time I had certain guidelines to follow and was told how much business I was expected to handle. I spent most of the time on my own, travelling to farms to discuss sales contracts for grain to be harvested the following year and then in some

cases selling it to other commercial outlets. In order to keep up with sales targets and grain requirements I worked long hours, usually nine- to ten-hour days.

After successfully handling this line of work I was appointed grain trader at one of our regional offices. This meant I had full responsibility for the trading of feed and malting barley in the area.

My latest promotion came last year, when I was offered the post of branch trading manager. I now find I spend an increasing amount of time in the office conducting trading in grain and arable products over the telephone, relying on other buyers to check grain quality on the farm. I do still visit farms and mills but not to the same extent as during my training, and planning, writing, conferring and advising are now the dominant features of my work, along with some market research. Although I do not like being less involved in the market at farm level, I do enjoy the feeling of being responsible for a job well done.

My career structure is flexible and depends on my personal achievements. Promotion prospects are, as you can see, good. I do not feel limited within this company, as there are many different trading departments I could transfer to.

Agricultural Marketing

The purpose of the agricultural marketer is to identify customer needs, supply the appropriate products, making sure they are of the highest quality, and provide a competitive price. This involves financial and market planning and needs an up-to-date knowledge of the market-place and the current buying trends.

The usual method of entry into this field is either by taking a business studies course at college or polytechnic, or by sitting for a degree at university. In some companies, sales departments will operate schemes, taking young people with a good secondary education (to A level standard), on to training courses that combine practical selling with a college course to produce salesmen. The current trend, though, does seem to be towards employing graduates, who will only need a short training period before being put out on the road. Entry to posts leading to careers in market and sales planning and management is usually at degree or BTEC higher diploma level. A large number of polytechnics, colleges and universities offer courses in business studies, but Newcastle University is at present the only place offering a degree specialising in agricultural economics and marketing.

For those who go into direct sales, the usual career structure is to begin as a trainee sales representative and go via sales representative and sales manager to a sales executive post. With promotion one moves away from selling directly to the customer and co-ordinates a junior sales force.

On the planning side, one may begin as a marketing assistant and move on through marketing manager to marketing executive.

The vast majority of vacancies in agricultural marketing occur in the agricultural supply industries, for example in machinery or fertiliser companies, or in the domestic food industries in either sales or planning advisory posts.

Case Study

David, at 24, is now a sales manager for a large food company and he specialises in dairy products. He is responsible to a management board, but takes most business decisions on his own and implements his own ideas to improve sales and those of his work-force.

> When I first joined the company, I wanted to train for a career that involved sales but also kept in touch with other people. Although I now have such a job, it is not what I originally envisaged, as my contact with people now comes from a personnel role and I have little direct customer contact.
>
> To begin with, the company made me a sales liaison assistant, where I worked with a more senior member of staff as a link between the customer and our dairy product lines. Most of my time was spent in the office ensuring the right products were delivered where and when they were needed and in the best condition. Much time was taken up with customer complaints or enquiries which I tended to investigate personally. As an assistant I was not in control of any staff, as I was there to learn about all aspects of the supply trade, from the bottom to the top. I found the training

challenging and rather demanding. One needs to be level-headed and have a lot of self-control; some customer complaints may be awkward to settle as one must take all aspects into account.

After both the company and I felt (by interview and progress reports) that I had mastered the essentials, I was promoted to area sales manager.

I am now directly responsible for the well-being of company employees. My sales area has various targets to be reached throughout the year and I must motivate my staff (of around 25) to promote our dairy products and expand our retail and domestic sales. Getting the best from my sales staff and achieving a team spirit is an area of the job which I enjoy greatly. I now spend more of my time outside visiting my staff, talking through problems, keeping them up to date with new lines and helping them increase sales. Back in the office I control the handling of complaints but usually leave a more junior member of staff to sort them out if possible. I do find, unfortunately, that we are not always able to answer customers' complaints to their satisfaction. My other function is to recognise gaps in the market and try, with the help of our research and development division, to produce new items for sale, as well as exploring other potential business areas.

I work fairly set hours, I suppose, between 9 and 6 pm, but like to remain flexible, according to staff and customer demands. I find, overall, that my attention is divided quite evenly between the different areas of the job.

I have a business studies sandwich degree from a polytechnic which was important in providing a base for me but my lack of knowledge of the dairy supply industry has been covered by the training I have since received. Prospects are good and I am more than satisfied with my gradual promotion towards senior sales management.

Vacancy Information

Farmers Weekly, the national press and college and university careers services are valuable sources of information on the current job market. Faculty notice-boards should also be checked regularly.

Chapter 6
Agricultural Engineering

The field of agricultural engineering has grown up because of the needs of farmers and growers to have solid engineering principles applied to their science. As farming in both the United Kingdom and overseas has become more intensive, and the need to grow more food more efficiently has expanded, mechanisation and automation have come to play an increasingly vital role. In addition, current advances in research and improved techniques of farming and harvesting have led to a demand for ever more sophisticated machinery, so allowing the whole farming cycle to be carried out more effectively.

Agricultural engineering has become a highly specialised field but at the same time overlaps with many other engineering subjects and is an interdisciplinary study. Integration is thus a key word in the training of agricultural engineers, who need substantial theoretical and practical knowledge of both agriculture and engineering but not as separate disciplines. However, it is still possible to follow a career in this field even when starting in a single related engineering discipline, such as civil engineering or chemical engineering, or by taking an agricultural course, although the student will eventually have to combine the two on a postgraduate scheme.

The three main levels of engineer in the industry are now looked at in some detail, along with an examination of the type of career prospects they can look forward to and a brief summary of the sort of places to find employment.

Agricultural Engineering

The Operator/Mechanic

The main function of the mechanic is the maintenance and repair of farm machinery. Although many farm workers will carry out simple repairs and keep equipment in good general running order, agricultural mechanics are taught at colleges to be specialists in the field.

As a rule, many qualified mechanics work for manufacturers, dealers and contractors of farm machinery. This is because only large farming units find it economical to employ full-time mechanics, having enough machinery needing constant servicing and maintenance to keep such a person fully employed.

Training usually begins for the young farm/company employee or apprentice with a BTEC qualification in agricultural mechanics, which has recently replaced the City and Guilds of London Institute examinations in many areas. This course provides a basic knowledge of agricultural machinery workshop processes and teaches maintenance and repair of farm implements and general equipment. The basic aim is to complement the practice gained by the trainee in his normal daily work and to introduce him to skills in the handling of machinery from a wider field than normally encountered.

The certificate aims to be an integrated course lasting three years. Most colleges and technical institutions make courses available on a full-time or block-release system, or allow the student to follow a day-release scheme. Many agricultural colleges run both their own and nationally controlled courses leading to certificates or diplomas.

Most colleges will just require students to have received a good general secondary education, but a few do prefer to take people who have studied preliminary engineering courses or had practical experience. To find out up-to-date requirements of individual colleges, prospective candidates should approach those colleges that are of interest.

More advanced training is available with some colleges providing additional specialist courses leading to their own diplomas. For the mechanic who becomes more academically inclined, Newcastle University offer a three-year first degree in agricultural mechanisation. General entry requirements are five GCE passes, including two at A level, or an acceptable technical qualification. Physics and mathematics would be needed at O level, and at A level chemistry or applied/additional maths would also be acceptable.

The Craftsman/Technician

As yet, no specific division exists in agriculture between the work of the technician and that of either the mechanic or the professional engineer. The main reason for this is the small size of the industry at present, which allows a degree of job flexibility not found elsewhere.

The majority of technicians are employed in field engineering, advisory work, manufacture, demonstration and technical writing and sales. A number also take up posts in management, distribution, installation of equipment — for example, field drainage systems, automated grain feeders, and grain dryers — or enter at foreman level. A few will find positions in design and development. Technicians can look forward to good career prospects with chances to advance to senior advisory or managerial posts following some years of practical experience and perhaps further study.

The general course of education for the technician begins with an ordinary national diploma in engineering, followed by a higher national diploma course. This leads to membership of the Institute of Agricultural Engineers, which for those who want promotion is essential. Many grades for membership exist for engineers of all levels and it is advisable to contact this Institute early in one's career.

The Professional Engineer

The primary concerns of the professional engineer in the farming industry lie in the research, design, development, production and marketing of agricultural machinery. There are three main branches of agricultural engineering:

Mechanical engineering, which provides the principles applied in the design, development and production of machinery and vehicles.

Civil engineering. This is relevant to the different areas involved in field engineering work. These include drainage, irrigation, soil mechanics, water supplies, and the transport route design of, for example, roads and the airstrips for light aircraft used in crop spraying. Civil engineers are more likely to find career opportunities overseas in both the developed and the less developed countries than engineers specialising in the other two branches.

Environmental control engineering. This is used in the planning, design and construction of farm buildings. These range from glasshouses and nursery buildings to milking sheds, poultry houses and crop-drying stores and barns.

Professional engineers may work in research and development departments of private companies or government departments at national and local level, for example within the Ministry of Agriculture. Other engineers become involved in advisory work in this country or abroad, and some take up technical sales and design posts with manufacturers. Career prospects are good and engineers can expect to be promoted to senior posts, for example to become chief engineers, project managers or senior technical advisers.

The level of professional engineer can be reached via technician training, but the more recognised route is now through a degree course.

Future Prospects

The examination system in agricultural engineering allows anyone to move up from one grade of membership to the next, within the Institute, provided one has the necessary practical experience and academic qualifications. No one need feel limited, therefore, at a particular level in the profession. It is also worth bearing in mind that a graduate is academically qualified for entry to the Council of Engineering Institutions and may, therefore, work in other branches of engineering.

Case Studies

A 26-year-old scientific officer who works in a research institute with 10 other engineers in his immediate department, describes his work.

> The particular branch I work for is chiefly concerned with overseas development and supplies agricultural engineers as project advisers.
>
> My colleagues and I divide our working time between the United Kingdom and overseas. When in this country I tend to work a set 8.30 am to 5 pm day and am concerned with ordering and despatching equipment to staff overseas and researching information on updated machinery. My usual place of work is the office, although I spend more time on paperwork here than I would like.
>
> It is the travel and experience and challenges encountered overseas that I really enjoy and I spend approximately one-third of the year abroad, visiting projects and advising on drainage schemes being planned in the field. I may be away for anything between two weeks and three months at one time. When overseas, I work no set hours and expect to put in 10-12 hour days, although a lot of time is spent travelling, either to sites or to collect supplies despatched from England.
>
> Since being here I have continued my education in two ways. I regularly attend short courses held at engineering institutes and colleges and I learn practical skills from more experienced members of the crew in the field. My initial qualification — a degree in agricultural engineering — has proved essential, however, and I have a farming background which has also been useful. Prospects for progress look good, with promotion to senior advisory posts possible when I have rather more experience, although I expect promotion to higher scientific officer later this year.

A 25-year-old project engineer gained a degree in engineering science and now works for a private food group that has extensive ties in the agricultural industry.

> On leaving university, I wanted to join a fairly large company that would allow me to continue my education and at the same time use the skills I already had on practical projects.
>
> When I started, I was given a brief induction course,

together with about 40 other newly appointed graduates, who were to go to other divisions as trainees. We were told about the structure of the company, the various divisions, company products, research and development being carried out and future career prospects. I then spent six months on a training course in basic workshop practice, followed by a further six-month course in mechanical engineering, which included work organisation, project engineering and business administration.

On the satisfactory completion of this training, I was promoted from trainee engineer to project engineer, which is the post I still hold. My present job means I work as part of a project team (all of whom are qualified engineers and technicians), and we have to satisfy the engineering requirements arising in other divisions and take on advisory work. We are currently involved in a variety of projects, examples being the design and construction of a new bulk storage facility for powdered food and a new processed-food production line. Although we are responsible to the chief engineer, nearly all production decisions are made by the group.

I tend to work an eight-hour day, five days a week, spending most of my time on research, planning and development. I get great satisfaction from being able to help design and put into practice a piece of equipment needed in the food division to improve product quality and production.

My career possibilities are promising, as the company takes care to give its employees a comprehensive training; I am still learning and expanding my knowledge and do not feel restricted in my work. I now have experience in production, maintenance and general administration, all of which provide a good base on which to develop my future. I would also say that my degree has provided invaluable background knowledge, as well as being of prime importance in getting offered the job.

Vacancy Information

General vacancies may be found in the local press, or *Farmers Weekly*. Those with technical and professional qualifications should look in the above as well as in the national press, in university careers service job sheets, on faculty notice-boards and departmental boards. Those wishing to pursue a career in research should also look in *New Scientist*.

Chapter 7
Other Careers Available

In this chapter a number of additional positions are explored briefly. The fact that they are not being given as much space as previous subjects does not make them any less important as career lines. They are looked at quickly because they fall into areas where it becomes more difficult to draw a line between agriculture and its sciences and other fields. These interrelated jobs do, however, all deserve to be brought to the reader's attention.

Fish Farming

Fish farming is a developing industry, offering more and more job opportunities. Trout, which are now farmed in increasing numbers, are no longer considered a luxury food and the industry now resembles other intensive farming operations, like the poultry industry, as it concentrates on intensive stock rearing.

Fish farming supplies trout and carp, for example, of uniform size and quality, in response to a strong demand from the catering trade, hotels, restaurants and pubs, specialised trades and the domestic market. Existing farms now rear fish in their thousands and therefore need to apply proper husbandry and management techniques to ensure the healthy rearing of their product.

Opportunities occur at all levels from assistant to manager, with the chance for a limited number, who have the right financial and technical backing, to set up their own farms. A spin-off is the need for fisheries advisers to instruct on correct feeding, breeding and hygiene. Such posts are available in Ministry of Agriculture departments and some agricultural supply companies.

Full information on this subject can be obtained from the Ministry of Agriculture, and details of courses run in fish farming are given by the Hampshire College of Agriculture.

Forestry

Forestry is both a science and an art, providing many varied career opportunities. It can be divided into the categories of forest worker, forester and forest officer, which are covered below.

The largest employer in the United Kingdom is the state-owned Forestry Commission, with other posts available in timber and nursery organisations in the private sector.

The Forest Worker

The nature of this position demands the ability to operate machinery and carry out manual work. This involves many skills, including those of fencing, draining, planting young stock, timber harvesting and nursery work. To acquire these varied skills, the forest worker undergoes an integrated course of study, involving theoretical and practical teaching. The aim of the training is to bring the worker up to craftsman status, which is desirable as it leads to prospects of promotion to ranger. Apprenticeships are not available but a forest worker is given encouragement and help to study for academic qualifications after gaining a post.

The City and Guilds of London Institute runs courses leading to forestry certificates, and full information is available from them.

The Forester

This job is that of technical manager. One is responsible for supervising and training workers, planning work programmes, financial costing, co-ordinating use of the forest by others for sport and recreation, improving and building up relations with local landowners on bordering land to their mutual benefit, as well as day-to-day running problems.

Educational qualifications for such a post are usually some kind of forestry certificate or diploma. Relevant courses are run by the Cumbria College of Agriculture and Forestry, Penrith.

The Forest Officer

Included in this area are the control of planting and conservation, together with forward planning and the overall control of staff out in the forest. The main route to entry at this level is to have a degree or higher diploma, although other qualifications combined with practical experience can be offered by the forester seeking promotion. Through further training and experience, promotion can be expected to forest officer grade 1 (this will take about five years) with future prospects then depending on individual ability and achievement.

Most of the top managerial posts are given to people with either a first degree or a postgraduate qualification in forestry. Competition for such posts is strong and people are well advised to offer as much practical experience and as good an academic record as possible.

Full details of career prospects and entry requirements can be obtained from the Chief Education and Training Officer, the Forestry Commission.

Overseas Work

This can be divided into agricultural work in the developed and less developed countries.

Developed Countries

With world unemployment ever increasing, it has become more difficult to obtain work overseas, with countries carefully regulating the influx of foreign workers, both seasonal and longer term, by tightening up their immigration laws. This makes it advisable to contact the agricultural adviser of the country or countries of your choice at the appropriate embassy or consulate in London, to enquire about work opportunities and conditions of entry.

Apart from joining a multi-national company in this country that has overseas postings, opportunities are few. When jobs do fall vacant they usually require specialised skills and/or experience. However, *Farmers Weekly* is an invaluable source of information on current prospects, both at home and abroad, and its 'situations vacant' columns will sometimes have foreign jobs advertised.

Less Developed Countries

An increasing number of young people are tempted by the thought of being able to travel overseas and help less able peoples to develop their land, but the reality is often very different from their initial expectations.

Voluntary Service Overseas, a major organisation sending volunteers out to work on specially designed projects, by specific request from developing countries, is totally honest about the hardships and pitfalls involved. Volunteers are sent out to projects to live and work as members of the community. They are paid a similar wage to the people they are assisting. They work abroad on a contract of set duration, which in some cases can leave a feeling of frustration at not being able to achieve as much as had been hoped in the time allowed.

The most important aspect to emphasise is that the Voluntary Service Overseas organisation needs skilled people. Volunteers with no practical skill or experience are of no use abroad as they are employed primarily to pass on skills and knowledge, by teaching and practical demonstration, to the local people who can then carry on work when the volunteer's service contract has expired. In the agricultural field, the service receives requests for general agriculturalists, livestock and crop specialists, foresters and agricultural engineers/mechanics.

If you are still interested in voluntary work despite the

known and recognised pitfalls — and for those people who do go on to be successful volunteers, there are many, many factors which make such work challenging and satisfying — the best course is to plan your education carefully and contact VSO, Oxfam or any of the other aid organisations as early as possible, to ask their advice on what is needed from you.

Part 2

Chapter 8
Courses and Qualifications

There are courses and qualifications at every level in agriculture, to suit the needs and abilities of practically anyone wishing to make a career in this field.

With so many different methods of study available, choosing the most suitable course requires you to know what you want to do eventually and how useful a particular scheme of study will be in helping you to find the right job and further your career. Before you finally decide which course to apply for, it is advisable to write to any of the places you are interested in, so you fully understand what the course entails, the entry qualifications required and any practical experience that is recommended.

The information given below covers all first degree courses offered by British universities in agriculture and the related sciences. Courses offered by agricultural colleges and other relevant institutions are listed in Chapters 9 and 10. Addresses are given in Chapter 11.

Details of part-time and evening courses offered by colleges of further education, technical colleges and polytechnics are not included, but such information is readily available from various government and professional bodies listed in Chapter 11. As course programmes change frequently, it is always advisable to check on the availability of a course before applying.

University First Degrees

These have been divided up according to subject matter with the agricultural sciences degrees grouped under the general headings of *botany* or *zoology* depending on the bias of the course.

As an independent, self-governing body, each university awards its own degrees and sets the conditions for entry to individual courses and the type of degree to be awarded.

In the lists given below, the notations used after the name of the university denote the type of degree awarded and are as follows: BA — Bachelor of Arts; BAgrHons — Bachelor of Agriculture with honours; BAHons — Bachelor of Arts with honours; BSc — Bachelor of Science; BSc(Hons) — Bachelor of Science with honours; BTech — Bachelor of Technology; Hons/Ord — a degree available at both honours and ordinary levels.

The number in brackets denotes the length in years of a particular course.

Agriculture
Aberdeen: BScHons/Ord (4)
Belfast: BAgrHons/Ord (3/4)
Edinburgh: BScHons/Ord (3/4)
London, Wye College: BScHons (3)
Newcastle: BScHons (3)
Nottingham: BScHons (3)
Reading: BScHons (3)
Wales, Aberystwyth: BScHons (3)
Wales, Bangor: BScHons (3)

Agricultural Bacteriology
Belfast: BAgrHons (4/5)

Agricultural Biochemistry
Nottingham: BScHons (3)
Wales, Aberystwyth: BScHons (3)

Agricultural Biochemistry and Nutrition
Newcastle: BScHons (3)

Agricultural Chemistry
Belfast: BAgrHons (4/5)
Glasgow: BScHons/Ord (3/4)
Leeds: BScHons (3/4)

Agricultural Economics
Aberdeen: BScHons (4)
Edinburgh: BScHons/Ord (4)
Exeter: BAHons (3)
Glasgow: BScHons/Ord (3/4)
London, Wye College: BScHons (3)
Manchester: BA(Econ) Hons/Ord (3)
Newcastle: BScHons (3)
Nottingham: BAHons (3)
Reading: BScHons (3)
Wales, Aberystwyth: BScHons (3)
Wales, Bangor: BScHons (3)

Agricultural Engineering
Cranfield: BSc (3)
Newcastle: BScHons (3/4)

Agricultural and Environmental Science
Newcastle: BScHons (3)

Agricultural and Food Marketing
Newcastle: BScHons (3)
Wales, Aberystwyth: BScHons (3)

Agricultural Mechanisation
Newcastle: BScHons (3)

Agricultural Science
Edinburgh: BScHons/Ord (3/4)
Leeds (Animal or Crop): BScHons (3/4)

Ecological Science
Edinburgh: BScHons/Ord (3/4)

Environmental Biology
Nottingham: BScHons (3)

Food Science
Nottingham: BScHons (3)

Forestry
Aberdeen: BScHons (4)
Wales, Bangor: BSc/BScHons (3)

Medicinal, Agricultural and Environmental Chemistry
Brunel: BSc/BTech (2/3)

Microbiology
Bath: BSc (4)

Nutrition
Nottingham: BScHons (3)

Soil Science
Aberdeen: BScHons (4)
Edinburgh: BScHons/Ord (3/4)
London, Wye College (with Plant Nutrition): BScHons (3)
Newcastle (with Land Resources): BScHons (3)
Nottingham: BScHons (3)
Reading: BScHons (3)
Wales, Bangor (with Biochemistry): BScHons (3)

Wood Science
Wales, Bangor: BSc/BScHons (3)

Degrees with a Botanical Bias

Agricultural Botany
Belfast: BAgrHons (4/5)
Glasgow: BScHons/Ord (3/4)
Leeds: BScHons (3/4)
Nottingham: BScHons (3)
Reading: BScHons (3)
Wales, Aberystwyth: BScHons (3)
Wales, Bangor: BScHons (3)

Applied Biology
Bath: BSc (4)

Crop Production
Bath: BSc (4)
Edinburgh: BScHons/Ord (3/4)

Mycology and Plant Pathology
Belfast: BAgrHons (4/5)

Plant Science
Aberdeen: BScHons (4)

Bath: BSc (4)
London, Wye College: BScHons (3)
Newcastle: BScHons (3)
Nottingham: BScHons (3)

Degrees with a Zoological Bias

Agricultural Zoology
Belfast: BAgrHons (4/5)
Glasgow: BScHons/Ord (3/4)
Leeds: BScHons (3/4)
Newcastle: BScHons (3)
Nottingham: BScHons (3)

Animal Nutrition
Glasgow: BScHons/Ord (3/4)
Leeds (with Animal Physiology): BScHons (3)

Animal Physiology and Ecology
Bath: BSc (4)

Animal Science
Aberdeen: BScHons (4)
Edinburgh (Production): BScHons/Ord (3/4)
London, Wye College: BScHons (3)
Nottingham: BScHons (3)

Physiology and Biochemistry of Farm Animals
Reading: BScHons (3)

CNAA Degrees

Apart from the universities, other higher education establishments award degrees. These Council for National Academic Awards (CNAA) degrees may be taken at selected polytechnics and technical colleges throughout the country. Entry requirements can be obtained from the relevant polytechnics, which are listed below;

Agriculture
Plymouth Polytechnic: BSc/BScHons (4)

Science
Portsmouth Polytechnic (Specialisation in Plant Science): BSc/BScHons (3)

Agricultural Technology
Wolverhampton Polytechnic: BSc (4)

CNAA Award

This is a postgraduate qualification that requires practical experience.

Diploma in Management Studies (Agricultural Option)
Seale-Hayne College (full time)

Higher Degrees and University Diplomas

The higher degrees available in universities in England, Scotland, Wales and Northern Ireland are listed below, under subject matter. Entry requirements to higher degrees usually specify a good first degree in an appropriate subject, but again, in some cases universities and other educational establishments will also consider similar qualifications together with practical experience. In agriculture, practical experience can, in certain cases, be just as important as academic experience. The prospective student should once again approach individual teaching establishments to clarify entry requirements.

General Agriculture
Higher degrees by research:
Aberdeen: MSc/PhD
East Anglia: MPhil/PhD
Edinburgh: MSc/MPhil
Glasgow: MSc/PhD
Leeds: MPhil/PhD
London, Wye College: MPhil/PhD
Newcastle: MSc/PhD
Nottingham: MPhil/PhD
Reading: MPhil/PhD

Wales, Aberystwyth: MSc/PhD
Wales, Bangor: MSc/PhD

Higher degrees by instruction:
London, Wye College: MSc Agriculture (one/two years)
Newcastle: MSc Agriculture
Reading: MSc Tropical Agricultural Development (one year); MSc/MAgrSc Tropical Agricultural Development (one/two years)
Wales, Aberystwyth: MSc Agricultural Science (one year)

Diplomas:
Aberdeen: DipLE Land Economy
Belfast: DipAgrComm Agricultural Communication
Edinburgh: Rural Science
Wales, Aberystwyth: Agricultural Science

Agricultural Botany, Agricultural Biology, Crop and Plant Science
Higher degrees by research:
Aberdeen: MSc/PhD
Bath: MSc/PhD
Cambridge: MPhil
Edinburgh: MSc/MPhil/PhD
Glasgow: MSc/PhD
Leeds: MPhil/PhD
Newcastle: MSc/PhD
Reading: MPhil/PhD
Wales, Aberystwyth: MSc/PhD
Wales, Bangor: MSc/PhD
Wales, Cardiff: MSc/PhD

Higher degrees by instruction:
Aberdeen: MSc Crop Science/Ecology (one year)
Brunel: MSc Weed Biology (one year)
Edinburgh: MSc (one year)
London, Wye College: MSc Applied Plant Sciences (one/two years)
Newcastle: MSc Agricultural Biochemistry and Nutrition; MSc Agricultural Biology

Reading: MSc Grassland Science; MSc Plant Taxonomy; MSc Crop Physiology; MSc Technology of Crop Protection
Wales, Aberystwyth: MSc Plant Breeding

Diplomas:
Aberdeen: Modern Botanical Methods; Forestry
East Anglia: Plant Biology
Edinburgh: Seed Technology
Wales, Bangor: Crop Protection

Agricultural Economics
Higher degrees by research:
Aberdeen: MSc/PhD
Belfast: MAgr/PhD
Cambridge: MPhil
Edinburgh: MSc/MPhil/PhD
Exeter: MSc/PhD
Glasgow: MPhil
Leeds: MPhil/PhD
London, Wye College: MPhil/PhD
Manchester: MA(Econ)/PhD
Newcastle: MSc/PhD
Oxford: MSc/MLitt/DPhil
Reading: MPhil/PhD
Wales, Aberystwyth: MSc(Econ)/PhD
Wales, Bangor: MSc/PhD

Higher degrees by instruction:
Aberdeen: MSc
Belfast: MAgr Farm Business Economics
Cambridge: MPhil Land Economy
Leeds: MA
London, Wye College: MSc Agricultural Economics; MSc Farm Business Management/Marketing
Manchester: MA(Econ)
Newcastle: MSc Agricultural Marketing; MSc Agricultural Economics
Oxford: MSc
Reading: MSc Agricultural Economics; MSc Agricultural Management

Wales, Aberystwyth: MSc Economics of Agricultural Production and Farm Management, Marketing and Agricultural Policy

Diplomas:
Aberdeen: Agricultural Economics
Newcastle: Agricultural Marketing
Reading: Agricultural Economics; Agricultural Methods
Wales, Aberystwyth: Agricultural Economics

Agricultural Sciences
Higher degrees by research:
Aberdeen: MSc/PhD
Birmingham: MSc/PhD
Cranfield: MSc/PhD
Glasgow: MSc/PhD
London, Imperial College: MPhil/PhD
London, Wye College: MPhil/PhD
Oxford: BLitt/MSc/DPhil
Wales, Bangor: MSc/PhD

Higher degrees by instruction:
Birmingham: MSc Conservation and Utilisation of Plant Genetic Resources
Cranfield: MSc Agricultural Engineering and Food Production
Reading: MSc Agricultural Meteorology

Diplomas:
Wales, Bangor: Agricultural Science

Agricultural Soil Science
Higher degrees by research:
Aberdeen: MSc/PhD
Cambridge: MPhil
Edinburgh: MSc/PhD/MPhil
Heriot-Watt: MSc/PhD
London, Wye College: MPhil/PhD
Newcastle: MSc/PhD
Nottingham: MPhil/PhD
Reading: MPhil/PhD

Wales, Aberystwyth: MSc/PhD
Wales, Bangor: MSc/PhD

Higher degrees by instruction:
Aberdeen: MSc
Cambridge: MPhil Soil Mechanics
Heriot-Watt: MSc Soil Mechanics
Newcastle: MSc Soil Science
Reading: MSc Pedology and Soil Survey and Soil Chemistry; MAgrSc Soil Science

Diplomas:
Aberdeen: Soil Science
Wales, Aberystwyth: Soil Science

Agricultural Zoology, Animal Science and Husbandry
Higher degrees by research:
Aberdeen: MSc/PhD
Belfast: MAgr/PhD
Glasgow: MSc/PhD
London, Wye College: MPhil/PhD
Newcastle: MSc/PhD
Wales, Bangor: MSc/PhD

Higher degrees by instruction:
Aberdeen: MSc Animal Nutrition; MSc Animal Production
Belfast: MAgr Dairy Technology; MAgr Poultry Technology
Edinburgh: MSc Animal Breeding
Newcastle: MSc Animal Nutrition
Nottingham: MSc Meat Science
Reading: MSc Animal Production; MSc Physiology and Biochemistry of Farm Animals
Wales, Bangor: MSc Animal Nutrition; MSc Animal Parasitology

Diplomas:
Aberdeen: Animal Nutrition
Edinburgh: Animal Breeding; Tropical Animal Production and Health; Animal Health

Liverpool: Bovine Reproduction
Wales, Bangor: Animal Nutrition; Animal Parasitology

Agricultural Engineering
Higher degrees by research:
Cranfield: MSc/PhD
Newcastle: MSc/PhD

Higher degrees by instruction:
Cranfield: MSc Agricultural Engineering and Food Production
Newcastle: MSc
Reading: MSc Technology of Crop Protection

Diplomas:
Belfast, Loughry College of Agriculture and Food Technology: Agricultural Communication
Belfast, National College of Agricultural Engineering: Postgraduate Diploma

For those interested in graduate studies in agricultural engineering at the Cranfield Institute, teaching is actually carried out at Silsoe College, Silsoe, and one should contact the Student Recruitment Officer there.

Chapter 9
Diplomas and Certificates in England and Wales

This chapter provides information on the ordinary and higher national diplomas and certificates available, as well as the awards of the Business and Technician Education Council (BTEC), the Welsh Joint Education Committee and the West Midlands Advisory Council. There are also details of regional and individual college-based qualifications, and the certificates offered by various societies and associations. Addresses are given in Chapter 11.

Colleges all over the country provide courses leading to the examinations of the City and Guilds of London Institute, offering the chance to acquire both basic and advanced skills in most forms of agriculture. The CGLI produce schemes of technical education, set examinations and establish national standards of expertise. For information on courses and colleges, contact the City and Guilds of London Institute at 76 Portland Place, London W1N 4AA.

Key to Abbreviations

AC	Agricultural College
AFEC	Agricultural Further Education Centre
AI	Agricultural Institute
ATC	Agricultural and Technical College
B	Block release
BTEC	Diploma or certificate offered by the Business and Technician Education Council
C	College
CA	College of Agriculture
CAF	College of Agriculture and Forestry
CAFT	College of Agriculture and Food Technology

CAH	College of Agriculture and Horticulture
CAT	College of Agriculture and Technology
CFE	College of Further Education
CFEA	College of Further Education and Agriculture
CFESA	College of Further Education and School of Art
CFHE	College of Further and Higher Education
CT	College of Technology
DR	Day release
E	Evening
FT	Full time
HNC	Higher National Certificate
HND	Higher National Diploma
IHE	Institute of Higher Education
OND	Ordinary National Diploma
TAC	Technical and Agricultural College
TC	Technical College
TCSA	Technical College and School of Art
S	Sandwich course

Business and Technician Education Council Awards

The entry requirement for BTEC national certificates and diplomas is normally a good education to O level standard to include one or more O levels in science subjects. For the higher certificates, one requires formal agricultural qualifications, eg the BTEC national certificate or equivalent plus practical experience in some cases.

Higher National Diplomas

As is the case with the ordinary national diploma, the HND is a three-year sandwich course and candidates for any of the agricultural colleges must have spent at least one year working on a farm or in a similar acceptable establishment. The HND in agriculture requires students to have at least one A level in a science subject, with supporting O levels in appropriate areas. An OND or ONC qualification may also be acceptable. These courses are intended for people who will be involved with the managerial and technical/technological aspects of farming.

Careers in Agriculture

General Agriculture
Harper Adams AC; Lancashire CAH; Royal AC; Seale-Hayne AC; Shuttleworth AC (with arable farming option); Welsh AC; Writtle AC

Agricultural Engineering
Harper Adams AC (S); Rycotewood C (FT) (BTEC); Wolverhampton Polytechnic (S)

Agricultural Marketing and Business Administration
Bishop Burton CA (S); Harper Adams AC (S)

Fishery Management
Hampshire CA

Mechanical Engineering (Agriculture)
Chelmer IHE (S); Writtle AC (S)

Rural Resources and Their Management
Seale-Hayne AC

Supplementary Farm Management
Seale-Hayne AC (FT)

Ordinary National Diplomas

For entry to OND courses, students must normally have some practical experience (usually one year), together with GCE O level or CSE grade 1 passes in four subjects, to include suitable science subjects. Individual colleges should be approached to confirm the subjects needed, and also for help and guidance in finding suitable practical work.

General Agriculture
Berkshire CA; Bicton CA; Bishop Burton CA; Brooksby AC; Cheshire CA; Derbyshire CA; Durham AC; Gloucestershire CA; Hadlow CAH; Hampshire CA; Hertfordshire CA; Lancashire CA; Lincolnshire CAH (with arable bias); Norfolk CAH; Northumberland CA; Plumpton AC; Shuttleworth CA; Somerset CAH; Walford CA; Warwickshire CA; Welsh CA; Writtle AC

Agricultural Engineering
Carmarthen CAT (BTEC); Cheshire CA (BTEC);

Lackham CA (BTEC); Lincolnshire CAH (BTEC);
Rycotewood C (BTEC); Somerset CAT (BTEC)

Agricultural Merchanting
Berkshire CA (S); Nottinghamshire CA (FT)

Agricultural Science
Carmarthen CAT (BTEC) two-year full-time course

Agricultural Subjects
Berkshire CA (S) (farm mechanisation bias; pig production bias); Lackham CA (S) (farm mechanisation bias); Northamptonshire CA (livestock farming bias); Plumpton CA (livestock farming bias); The Usk CA (livestock farming bias); Welsh AC (supplementary farm business organisation and management)

Business Studies: Agricultural Secretary
These are full-time courses lasting two years.
Aylesbury C (BTEC); Bishop Burton CA (BTEC); Brooksby AC (BTEC); Hampshire CA (BTEC); Lincolnshire CAH (BTEC)

Dairying
The courses offered at both colleges are a combination of Food Technology and Dairy Technology.
Cheshire CA; Seale-Hayne AC

Fishery Management
Hampshire AC

Forestry
Cumbria CAF (S); Merrist Wood AC (S) (arboriculture)

Horse Management
Warwickshire CA

Poultry Husbandry
Harper Adams AC

National Diploma in Agriculture
Students are expected to be of O level education standard, offering one or more relevant subjects plus at least a year's practical experience.
Bicton CA

Ordinary National Certificates

The entry requirement for these courses is a good general education (considered to be up to GCE O level or CSE grade 1 standard), including English, mathematics and a science subject, preferably followed up by a part-time course in agriculture or horticulture before starting.

Students must normally have worked for at least one year on the land after leaving school. The national certificates themselves are taken at the end of one-year full-time courses and the students benefiting most from them are usually those intending to become practical farmers or growers.

Agriculture
Askham Bryan CAH; Aylesbury CFEA; Berkshire CA; Bicton CA; Bishop Burton CA; Brooksby AC; Carmarthen TAC; Chadacre AI; Cheshire CA; Cumbria CAF; Derbyshire CA; Dorset CA; Durham AC; Gloucestershire CA; Hadlow CAH; Hampshire CA (also with dairy farm bias); Hertfordshire CA; Lackham CA; Lancashire CA; Lincolnshire CA; Llysfasi CA; Merrist Wood AC; Norfolk CAH; Northampton CA; Northumberland CA; Nottinghamshire CA; Plumpton AC; Somerset CAH; Staffordshire CA; The Usk CA; Walford CA; Warwickshire CA; Writtle AC

Agriculture with Home Economics
Lackham CA; Llysfasi CA; Northampton CA; Nottinghamshire CA; Staffordshire CA; The Usk CA

Agriculture: Hill and Upland Farming
Cumbria CAF; Llysfasi CA

Agriculture with Dairy Farm Bias
Hampshire CA

Agriculture with Supplementary Advanced Husbandry
Durham AC

Agriculture with Supplementary Certificate in Arable Farm Management and Mechanisation
Lincolnshire CA

Agriculture with Supplementary Dairy Herd Management
Cheshire CA

Agriculture with Supplementary Large Livestock Production
Northampton CA

Business Studies
North Tyneside CFE (equestrian option)

Dairying
Cheshire CA (BTEC) dairy technology; Somerset CAH

National Certificate in Farm Management (full time)
Bishop Burton CA; Durham AC; Hampshire CA; Merrist Wood AC

National Certificate for Farm Secretaries
Aylesbury CFEA; Bicton CA; Brooksby AC; Hadlow CA; Hampshire CA; Lincolnshire CAH; Northumberland CA; Staffordshire CA; The Usk CA

Veterinary Science
Brooklands TC

National Stockman's Certificate in Poultry Practice
A good level of secondary education is required, but the course carries no specific academic regulations.
Lincolnshire CA; Plumpton AC

Post National Stockman's Certificate in Poultry Practice
Lincolnshire CA

Advanced Certificates

The entry requirement for these advanced national certificate courses is a good standard in the ordinary national certificate course, or in a comparable examination. It is possible to take some immediately after completion of the national certificate, but for others (especially those involving management) students should have spent some time in a post of responsibility following their basic course.

The courses are designed to cover in greater depth the

various specialised branches of agriculture and horticulture. They may last up to nine months (a full academic year).

Agriculture
Berkshire CA; Bicton CA (also with farm machinery and mechanisation option); Warwickshire CA

Agricultural Engineering
Lackham CA (FT) (BTEC); Lincolnshire CAT (FT) (BTEC)

Dairy Farm Management
Derbyshire CA; The Usk CA

Dairy Herd Management
Cheshire CA

Farm and Grassland Management
Berkshire CA; Bicton CA (including a machinery-mechanisation option)

Farm Management
Bishop Burton CA; Brooksby AC; Hampshire CA (course in livestock on the arable farm); Welsh AC (supplementary certificate in farm business organisation)

Farm Management and Mechanisation
Lincolnshire CA (course includes arable and general options)

Livestock Management
Lackham CA

Livestock Management with Dairy, Sheep and Beef
Chippenham TC

Livestock Production and Business Management
Northamptonshire CA

Lowland Sheep Production
Hampshire CA

Maintenance and Repair of Agricultural Machinery
Warwickshire CA

Management on the Intensively Grazed Mixed Farm
Warwickshire CA

Pig Husbandry
Norfolk CAH

Pig Unit Management
Bishop Burton CA

Science (supplementary unit in crop production)
Norwich City CFHE (BTEC)

Sheep Management
Hampshire CA; Northumberland CA

Welsh Joint Education Committee Awards

Students wishing to enter Phase I courses should have up to five O levels; for higher phases, practical experience is needed.

Agriculture Phase I
Brecon CFE (DR); Carmarthen ATC (DR); Chester CFE (DR); Glynllifon C (FT; DR); Llysfasi CA (DR); Meirionnydd C (DR); Mid-Glamorgan CAH (DR); Montgomery CFE (FT; DR); Pembrokeshire TC (DR); Pencraig C (DR); The Usk CA (DR)

Agriculture Phase II
Glynllifon C (FT; DR)

Agriculture Phase II (Grassland and Forage Crops)
Brecon CFE (DR); Glynllifon C (DR); Meirionnydd C (DR); Mid-Glamorgan CAH (DR); Pembrokeshire TC (DR); Pencraig C (DR)

Agriculture Phase II (Beef Production)
Brecon CFE (DR); Carmarthen ATC (DR; E); Glynllifon C (DR); Llysfasi CA (DR); Meirionnydd C (DR); Mid-Glamorgan CAH (DR); Montgomery CFE (DR); Pembrokeshire TC (DR); Pencraig C (DR); The Usk C (DR)

Agriculture Phase II (Hill Farming)
Brecon CFE (DR); Merionnydd C (DR); Montgomery CFE (DR)

Agriculture Phase II (Sheep Production)
Brecon CFE (DR); Carmarthen ATC (DR; E); Glynllifon C (DR); Llysfasi CA (DR); Merionnydd C (DR); Mid-Glamorgan CAH (DR); Montgomery CFE (DR); Pencraig C (DR); The Usk CA (DR)

Agriculture Phase II (Milk Production)
Brecon CFE (DR); Carmarthen ATC (DR; E); Glynllifon C (DR); Llysfasi CA (DR); Mid-Glamorgan CAH (DR); Montgomery CFE (FT; DR); Pembrokeshire TC (DR); The Usk CA (DR)

Agriculture Phase II (Farm Machinery)
Brecon CFE (DR); Carmarthen TAC (DR; E); Glynllifon C (DR); Mid-Glamorgan CAH (DR); Montgomery CFE (DR); Pembrokeshire TC (DR); Pencraig C (DR); The Usk CA (DR)

Agriculture and Horticulture, Phase III (Accounts)
Glynllifon C (DR); Llysfasi CA (DR)

Farm Enterprise Management Phase III
Pembrokeshire TC (DR)

Farm Enterprise Management Phase IV
Pembrokeshire TC (DR)

West Midlands Advisory Council Awards

These courses require students to have a sound secondary education plus, in some cases, practical experience.

Agriculture Phase I
Bridgnorth CFE (DR); Canterbury CT (DR); Gloucestershire CA (DR); Hereford CA (FT; DR); Highlands C (FT); Leek CFE (DR); Merrist Wood AC (DR); Somerset CA (DR); Staffordshire CA (DR); Warwickshire CA (DR); Worcestershire CA (DR)

Agriculture Phase II
Warwickshire CA (DR)

Agriculture Phase II (Animal Production)
Bridgnorth CFE (DR); Gloucestershire CA (DR);

Hereford CA (FT; DR); Merrist Wood AC (B);
Worcestershire CA (DR)

Agriculture Phase II (Beef Production, Advanced)
Hereford CA (DR); Merrist Wood AC (DR);
Warwickshire CA (B); Worcestershire CA (DR)

Agriculture Phase II (Sheep Production, Advanced)
Hereford CA (DR); Merrist Wood CA (DR);
Warwickshire CA (B); Worcestershire CA (DR)

Agriculture Phase II (Tractor Operation)
Hereford CA (DR); Warwickshire CA (B);
Worcestershire CA (DR)

Agriculture Phase II (Farm Workshop Practice)
Worcestershire CA (DR)

Agriculture Phase II (Mechanised Crop Production, Intermediate)
Bridgnorth CFE (DR); Gloucestershire CA (DR);
Hereford CA (FT; DR); Somerset CA (DR);
Worcestershire CA (DR)

Agriculture Phase II (Crop Production, Advanced)
Hereford CA (DR); Warwickshire CA (B);
Worcestershire CA (DR)

Agriculture Phase II (Grassland and Fodder Crops)
Gloucestershire CA (DR); Hereford CA (DR);
Merrist Wood AC (DR); Warwickshire CA (B);
Worcestershire CA (DR)

Agriculture Phase II (Advanced Cash Roots – Potatoes and Sugar Beet)
Hereford CA (DR); Warwickshire CA (B)

Agriculture Phase II (Advanced Pig Production)
Merrist Wood AC (DR)

Agriculture Phase II (Advanced Milk Production)
Gloucestershire CA (DR); Hereford CA (DR);
Merrist Wood AC (DR); Warwickshire CA (B);
Worcestershire CA (DR)

Agriculture Phase III (Farm Records and Accounts)
Gloucestershire CA (DR); Hereford CA (DR); Merrist
Wood CA (E); Mid-Glamorgan CAH (DR); The Usk CA (E);
Worcestershire CA (DR)

Agriculture Phase III (Farm Enterprise Management)
Hereford CA (DR)

Agriculture Phase IV (Farm Business Management)
Gloucestershire CA (DR); Hereford CA (DR);
Hertfordshire CAH (E); Mid-Glamorgan CAH (DR);
The Usk CA (E); Warwickshire CA (FT; E); Worcestershire
CA (DR)

Awards from Other Bodies

Educational requirements vary, but application should be made to individual colleges for confirmation. Education should be to O level or CSE grade 1 standard, to include science subjects, and combined with a year's work experience.

British Bee Keepers Association

Preliminary Certificate in Bee Keeping
Bicton CA; Hampshire CA

British Horse Society

Certificate of Horsemastership
Chippenham TC; East Devon CFE

Assistant Instructor's Certificate
Carmarthen TAC; Chippenham TC; East Devon CFE;
Newark TC

Teaching Certificate
Newark TC

Central Forestry Examination Board

National Diploma in Forestry
Cumbria CAF (B); this course requires practical experience.

National Examinations Board for Supervisory Studies Awards

Certificate (Agriculture and Horticulture)
Cheshire CA (DR); Hampshire CA (FT); Stoke TC (DR); Welsh CH (FT); practical experience and a general secondary education are required for this course.

Manpower Services Commission Training Division

TOPS Course in Agricultural Machinery Repair
Herefordshire TC (FT)

Road Transport ITB

Integrated Training Course for Agricultural Machinery Apprentices
Stafford CFE (FT)

Royal Association of British Dairy Farmers

National Certificate in Dairying (full time)
Carmarthen TAC; Hampshire CA; Llysfasi CA; The Usk CA

Regional Awards

Agricultural Craft Practice
Bolton MetC (DR); Bridgnorth CFE (DR)

Agricultural Operatives, Craft Course – Special Options
Askham Bryan CAH (DR); Bedale AC (DR); Easingwold AC (DR); Guisborough AC (DR); Pickering AC (DR)

Agriculture Phase I
Bishop Burton CA (DR); Burton upon Trent TC (DR); Lancashire CA (B; DR); Wakefield District (DR)

Preliminary OND
Durham AC (B; DR); Easingwold AC (DR); Lincolnshire CAGH (DR); North Oxfordshire TCSA (FT)

Agriculture Phase II (Dairying Milk Production)
Gloucestershire CA (DR)

Careers in Agriculture

Advanced Certificate in Agricultural Merchanting
Nottinghamshire CA (FT)

College Awards

Students are normally required to have a general secondary education, up to O level standard, for entry to these courses. Practical experience may also be needed in some cases. (S) denotes sandwich courses.

Certificate in Agriculture
Askham Bryan CAH (FT); Lincolnshire CAH (FT); Royal AC (FT); Shuttleworth AC (FT)

Diploma in Agriculture
Bishop Burton CA (FT; S); Royal AC (FT)

Certificate in Animal Care
North East Surrey CT (FT)

Certificate in Gamekeeping, Waterkeeping and Fish Farming
Hampshire CA (FT)

Certificate in Pig Management
Bishop Burton CA (DR)

Advanced Certificate in Pig Management
Bishop Burton CA (FT)

Certificate in Poultry Husbandry
Lincolnshire ACH (S)

Certificate in Poultry Practice
Plumpton AC (FT)

Diploma in Fishery Management
Hampshire AC (FT)

Diploma in Hill and Marginal Farming
Cumbria CAF (S)

Diploma in Poultry Husbandry
Harper Adams CA (S)

Diploma in Poultry Management
Lincolnshire ACH (S)

Technician Diploma in Dairy Farm Management
Derbyshire CA (FT)

Certificate in Farm Management
Bishop Burton CA (FT)

Farm Recording and Secretarial Certificate
Staffordshire CA (S)

Diploma in Farm Management
Northumberland CA (S)

Diploma in Advanced Farm Management
Royal AC (FT)

Certificate in Agriculture, Machinery Option
Bicton CA (FT)

Certificate in Farm Mechanics
Hampshire CA (FT)

Certificate in Farm Mechanisation and Machinery Maintenance
Lincolnshire CAH (S)

Certificate in the Maintenance and Repair of Agricultural Machinery
Warwickshire CA (FT)

Certificate in the Maintenance and Repair of Arable Machinery
Bishop Burton CA (FT)

Diploma in Agricultural Engineering
Lincolnshire CAH (S)

Diploma in Arable Farm Mechanisation and Management
Lincolnshire CAH (FT)

College Based Awards

Pre-University Course in Practical Agriculture
Northampton CA (FT)

Careers in Agriculture

Preparatory Course for the HND in Agricultural Subjects
Harper Adams CA (FT)

Preliminary/Introductory Course in Agriculture
Aylesbury C (FT); Bedford CFE (FT); Bishop Burton
CA (DR); Derbyshire CA (FT); Hereford CA (FT);
Llysfasi CA (FT); Mid-Glamorgan CAH (FT);
Montgomery CFE (FT); Norfolk CAH (FT); Rockingham
CFE (FT); Solihull CT (FT); West Sussex CAH (FT)

Work Skills (Rural Industries)
Guisborough AC (S)

Meat Rabbit Production
Derbyshire CA (B)

Supplementary Course in Advanced Pig Husbandry
The Usk CA (S)

Courses in Bee Keeping/Hive Equipment Making
Writtle AC (E)

College Based Courses

A general secondary education is the only requirement for these courses.

Pre-Apprenticeship Course in Agricultural Engineering
East Devon CFE (FT)

Agricultural Hydraulics
North Nottinghamshire CFE (E)

Rural Engineering
Herefordshire TC (FT)

Agricultural Welding
Frome C (DR)

Arc Welding
Plumpton CA (E)

Oxy-Acetylene Welding
Plumpton CA (E)

Agricultural Secretaries (full time)
Aylesbury C; Bridgwater C; Chichester CT;
Herefordshire TC; West Kent CFE

Introductory Course for Agricultural Secretaries with Rural Economics
Hereford AC (FT)

Chapter 10
Diplomas and Certificates in Scotland and Northern Ireland

Scotland has a similar pattern of qualifications, and similar teaching establishments, to those found in England and Wales. It does not, however, provide the equivalent of farm institutes, so students at this level either take a course leading to a college certificate, or study part time for the City and Guilds of London Institute's exams, while they are working on an approved farm under the Scottish National Apprenticeships scheme.

Scotland has three central institutions which specialise in agriculture: East of Scotland College of Agriculture, Edinburgh; North of Scotland College of Agriculture, Aberdeen; West of Scotland College of Agriculture, Auchincruive, Ayr. The various courses offered by these colleges and other educational establishments are given below. All full postal addresses can be found in Chapter 11. Abbreviations used are explained on pages 66 and 67.

In addition to the courses listed here, the City and Guilds of London Institute conduct a wide range of examinations in Scotland and Northern Ireland.

Qualifications available at colleges in Northern Ireland are the same as those available elsewhere in the British Isles.

Scottish Technical Education Council Awards

The qualifications listed below can be attained on a part- or full-time basis, or as sandwich courses. The length of courses will thus vary as certain units are obligatory and a minimum number of course hours have to be completed. The majority can be completed in one year, but individual colleges should be approached to discuss how a course is taught there.

Basic entry requirements are one or more O level grades in appropriate science subjects for ordinary certificate and diploma courses and one or more A or H level sciences with three or four supporting O levels for higher diploma and certificate courses. Some practical work experience may also be asked for. It is anticipated that in mid-1985, SCOTEC will amalgamate with the Scottish Business Education Council (SCOTBEC) to form the Scottish Vocational Education Council (SCOTVEC).

Certificate in General Agriculture
Clinterty AC (FT); Elmwood ATC (FT)

Certificate in Agriculture
Borders AC (DR); Falkirk CT (DR); Kilmarnock C (DR); Motherwell C (DR); Oatridge AC (FT; B; DR); Perth CFE (DR); Reid Kerr C (DR); Thurso TC (DR)

Diploma in Agriculture
East of Scotland AC (S); Elmwood ATC (FT); North of Scotland AC (FT); Oatridge AC (FT); West of Scotland CA (S)

Higher Diploma in Agriculture
East of Scotland CA (S); North of Scotland CA (FT); West of Scotland CA (S)

Certificate in Agriculture Part I
Angus TC (DR); Barony AC (DR); Clinterty AC (DR); Elmwood ATC (FT; B; DR); Henderson TC (DR); Stranraer AFEC (DR)

Certificate in Agriculture Part II
Angus TC (DR); Barony TC (DR); Clinterty AC (DR); Elmwood ATC (B; DR); Henderson TC (DR); Stranraer AFEC (DR); Thurso TC (DR)

Certificate in Agriculture Part III
Barony AC (DR); Clinterty AC (DR); Elmwood ATC (B; DR); Henderson TC (DR); Stranraer AFEC (DR); Thurso TC (DR)

Elmwood and Clinterty Colleges also offer Part II agricultural options in the following areas: Cereal Production; Grassland and Fodder Crop Production; Hill and Upland Farming (Elmwood only); Potato Production (Elmwood only); Beef Production; Sheep Production; Pig Production (Elmwood only); Milk Production; Welding and Workshop Production (Clinterty only) and a Tractor and Machinery Option.

Certificate in Agricultural Engineering
Barony AC (B; FT); Clinterty AC (B; FT; S);
Elmwood ATC (B; FT); Oatridge AC (B; FT)

Higher Certificate in Agricultural Engineering
Elmwood ATC (FT)

Certificate in Farm Management
Barony AC (DR); Borders AC (DR); Elmwood ATC
(B; DR; FT); Kirkwall FEC (B; DR); Oatridge AC (B; DR);
Thurso TC (DR)

Diploma in Food Technology (Dairy Technology)
West of Scotland CA (S)

Higher Diploma in Food Technology (Dairy Technology)
West of Scotland CA (S)

Certificate in Poultry Husbandry
West of Scotland CA (FT)

Diploma in Poultry Production
West of Scotland CA (S)

Certificate in Forestry Part I
Barony AC (B); Inverness TC (B)

Certificate in Forestry Part II
Inverness TC (B)

Certificate in Forestry Part III
Inverness TC (B)

Diploma in Forestry
Inverness TC (FT)

Scottish National Diploma for Agricultural Secretaries
Elmwood ATC (FT)

National Certificates – Scotland

These are the same as those awarded in England and Wales and have similar academic requirements for entry. Courses are taken on a full- or part-time basis.

Agriculture
Clinterty AC (FT)

Farm Management
Elmwood ATC

College Based Awards – Scotland

The courses listed below are run by the individual colleges which award their own certificate to successful students on completion of the subjects. Courses vary in length between one and three years. Although no specific academic qualifications are asked for, it is assumed that students will be of O level standard in the sciences. A few colleges may wish candidates to have practical experience, so this should be confirmed by contacting the individual colleges.

Preliminary/Introductory Course in Agriculture
Elmwood ATC (FT)

Certificate in Bee Keeping
West of Scotland CA (DR)

Farm Records and Accounts
Thurso TC (DR)

Fish Farming
Barony AC (FT)

Highlands and Islands Development Board – Course in Fish Farming
Inverness TC (B)

Supplementary Certificate in Cheese and Whey Technology
West of Scotland CA (FT)

Careers in Agriculture

Supplementary Certificate in Milk Production Technology and Control
West of Scotland CA (FT)

Awards from Other Bodies

Manpower Services Commission

A full-time training course is available at Motherwell College under this scheme, with no formal educational requirements specified. Further details can be found by contacting the College or the MSC.

National Examinations Board for Supervisory Studies

Certificate in Forestry
Inverness TC (B)

National Diplomas — Northern Ireland

Ordinary national diploma courses require students to have up to five O levels, to include science subjects and maths. Higher level diplomas require one or two A levels as well as up to four supporting O levels in sciences.

Agricultural Sciences — General Agriculture
Greenmount CAH (S)

Food Technology (Dairy Technology)
Loughry CAFT (FT; S) (BTEC)

Higher National Diplomas — Northern Ireland

Agricultural Subjects — Poultry Husbandry
Loughry CAFT (FT; S)

Food Technology (Dairy Technology)
Loughry CAFT (FT; S) (BTEC)

National Certificates — Northern Ireland

No academic requirements are specified, but students

should have a secondary education to O level standard. Some practical experience may be required.

National Certificate in Agriculture
Enniskillen AC (FT); Greenmount CAH (FT); Loughry CAFT (FT)

Advanced National Certificate in Agriculture
Greenmount CAH (FT)

College Awards — Northern Ireland

Entry requirements vary with colleges, but most assume students to have a general secondary education up to O level standard or CSE grade 1.

Certificate in Agriculture
Enniskillen AC (FT)

Certificate in General Agriculture
Loughry CAFT (FT)

Introductory Course in Agriculture
Omagh TC (FT)

No specialised courses are currently offered at colleges in Northern Ireland in Forestry or Secretarial and Farm Management.

Chapter 11
Useful Addresses

This chapter gives the full postal addresses of all the educational establishments mentioned on previous pages, together with the addresses of societies, institutes, associations and clubs that are all intimately involved in agriculture.

Universities

Initial enquiries to universities about courses offered should be addressed to the registrar, who will despatch relevant information, unless otherwise stated. Those interested in postgraduate courses should make this quite clear, when seeking further information.

University of Aberdeen
University Office, Regent Walk, Aberdeen AB9 1FA; 0224 40241.

University of Bath
Claverton Down, Bath BA2 7AY; 0225 61244.

The Queen's University of Belfast
University Road, Belfast BT2 7GZ; 0232 245891.

University of Birmingham
PO Box 363, Birmingham B15 2TT; 021-472 1301.

Brunel University
Uxbridge, Middlesex UB8 3PH; 0895 37188

University of Cambridge
Board of Graduate Studies, 4 Mill Lane, Cambridge CB2 1RZ; 0223 358933.

University of East Anglia
Norwich NR4 7TJ; 0603 56161.

University of Edinburgh
University Office, The Old College, South Bridge, Edinburgh EH8 9YL; 031-667 1011.

Useful Addresses

University of Exeter
Northcote House, Queen's Drive, Exeter EX4 4QJ; 0392 77911.

University of Glasgow
University Avenue, Glasgow G12 8QQ; 041-339 8855.

Heriot-Watt University
Chambers Street, Edinburgh EH1 1HX; 031-225 8432.

University of Leeds
Woodhouse Lane, Leeds LS2 9JT; 0532 431751.

University of Liverpool
PO Box 147, Liverpool LS9 3BX; 051-709 6022.

London University
(Letters should be sent to individual colleges.)

London, Imperial College of Science and Technology
London SW7 2AZ; 01-589 5111.

London, Wye College
Wye, Ashford, Kent TN25 5AH; 0233 812401.

University of Manchester
Oxford Road, Manchester M13 9PL; 061-273 3333.

University of Newcastle upon Tyne
9 Kensington Terrace, Newcastle upon Tyne NE1 7RU; 0632 328511.

University of Nottingham
University Park, Nottingham NG7 2RD; 0602 56101.

University of Oxford
University Offices, Wellington Square, Oxford OX1 2JD;
0865 56747.

University of Reading
Whiteknights, Reading RG6 2AH; 0734 875123.

University of Strathclyde
McCance Building, 16 Richmond Street, Glasgow G1 1XQ;
041-552 4400.

University College of North Wales, Bangor
Bangor, Gwynedd LL57 2DG; 0248 351151.

University College of Wales, Aberystwyth
Old College, King Street, Aberystwyth, Dyfed SY23 2AX;
0970 3177.

Polytechnics and Educational Institutions Offering Degrees

Cranfield Institute of Technology
Cranfield, Bedfordshire MK43 0AL; 0234 750111.

Plymouth Polytechnic
Drake Circus, Plymouth PL4 8AA; 0752 21312.

Portsmouth Polytechnic
Ravelin House, Museum Road, Portsmouth PO1 2QQ;
0705 827681.

The Polytechnic: Wolverhampton
Molineux Street, Wolverhampton WV1 1SB; 0902 710654.

Agricultural Colleges and Colleges of Further Education

England

Abingdon College of Further Education
Northcourt Road, Abingdon, Oxfordshire OX14 1NA

Airedale and Wharfedale College of Further Education
Calverley Lane, Horsforth, West Yorkshire LS18 4RQ

Askham Bryan College of Agriculture and Horticulture
Askham Bryan, York, North Yorkshire YO2 3PR

Aylesbury College of Further Education and Agriculture
Department of Agriculture and Horticulture, Hampden Hall, Stoke Mandeville HP22 5TB

Barnsley College of Technology
Church Street, Barnsley, South Yorkshire S70 2AN

Bedale Agricultural Centre
Benkhill Drive, Bedale, North Yorkshire DL8 2EA

Bedford College of Higher Education
Cauldwell Street, Bedford MK42 9AH

Berkshire College of Agriculture
Hall Place, Burchett's Green, Maidenhead SL6 6OR

Bicton College of Agriculture
East Budleigh, Budleigh Salterton, Devon EX9 7BY

Bishop Burton College of Agriculture
York Road, Bishop Burton, Beverley, Humberside HU17 8QG

Bridgnorth and Shropshire College of Further Education
Stourbridge Road, Bridgnorth, Salop WV15 6AL

Bridgwater College
Bath Road, Bridgwater, Somerset TA6 4PZ

Brooksby Agricultural College
Brooksby, Melton Mowbray, Leicestershire LE14 2LJ

Burnley College of Art and Technology
Ormerod Road, Burnley, Lancashire BB11 2RX

Burton upon Trent Technical College
Lichfield Street, Burton upon Trent, Staffordshire DE14 3RL

Cambridgeshire Farm College
Milton, Cambridge CB4 4DB

Canterbury College of Technology
New Dover Road, Canterbury, Kent CT1 3AJ

Chadacre Agricultural Institute
Chadacre, Shimpling, Bury St Edmunds, Suffolk IP29 4DU

Chelmer Institute of Higher Education
Victoria Road South, Chelmsford, Essex CM15 1LL

Cheshire College of Agriculture
Reaseheath, Nantwich CW5 6DE

Chester College of Further Education
Eaton Road, Handbridge, Chester CH4 7ER

Chichester College of Technology
Westgate Fields, Chichester, West Sussex PO19 1SB

Chippenham Technical College
Cocklebury Road, Chippenham, Wiltshire SN15 3QD

Coalville Technical College
Bridge Road, Coalville, Leicester LE6 2QR

Cornwall Technical College
Redruth, Cornwall TR15 3RD

Craven College
High Street, Skipton, North Yorkshire BD23 1JY

Cumbria College of Agriculture and Forestry
Newton Rigg, Penrith

Derbyshire College of Agriculture
Broomfield, Morley, Derby DE7 6DN

Dorset College of Agriculture
Kingston Maurward, Dorchester, Dorset DT2 8PY

Durham Agricultural College
Houghall, Durham DH1 3SG

Easingwold Agricultural Centre
Hambleton Way, York Road, Easingwold, North Yorkshire YO6 3EF

East Devon College of Further Education
Bolham Road, Tiverton, Devon EX16 6SH

East Suffolk College of Agriculture and Horticulture
Otley, Ipswich, Suffolk IP6 9EY

Careers in Agriculture

Exeter College
Hele Road, Exeter, Devon EX4 4JS

Frome College (Further Education)
Park Road, Frome, Somerset BA11 1EU

Gloucestershire College of Agriculture
Hartpury, Gloucester GL19 3BE

Guisborough Agricultural Centre
Avenue Place, Redcar Road, Guisborough, Cleveland TS14 6AX

Hadlow College of Agriculture and Horticulture
Hadlow, Tonbridge, Kent TN11 0AL

Hampshire College of Agriculture
Sparsholt, Winchester, Hampshire

Harper Adams Agricultural College
Newport, Salop TF10 8NB

Harrogate Agricultural Centre
c/o Great Yorkshire Show Centre, Site Office, Hookstone Wood Road, Harrogate HG1 4SL

Harrogate College of Further Education
Haywra Crescent, Harrogate, North Yorkshire HG1 5BE

Hereford College of Agriculture
Holme Lacy, Hereford HR2 6LL

Herefordshire Technical College
Folly Lane, Hereford HR1 1LS

Hertfordshire College of Agriculture and Horticulture
Oaklands, St Albans AL4 0JA

Highlands College
PO Box 142, St Saviour, Jersey

Hinckley College of Further Education
London Road, Hinckley, Leicestershire LE10 1HQ

Huddersfield Technical College
New North Road, Huddersfield, West Yorkshire HD1 5NN

Isle of Ely College of Further Education and Horticulture
Ramnoth Road, Wisbech PE13 2JE

Isle of Man College of Further Education
Homefield Road, Douglas, Isle of Man

Isle of Wight College of Arts and Technology
Medina Way, Newport, Isle of Wight PO30 5TA

Kendal College of Further Education
Milnthorpe Road, Kendal, Cumbria LA9 5AY

Useful Addresses

Kesteven Agricultural College
Caythorpe Court, Grantham, Lincolnshire NG32 3EP

Lackham College of Agriculture
Lacock, Chippenham, Wiltshire SN15 2NY

Lancashire College of Agriculture
Myerscough Hall, Bisborrow, Preston, Lancashire PR3 0RY

Lancaster and Morecombe College of Further Education
Morecombe Road, Lancaster, Lancashire LA1 2TY

Leek College of Further Education and School of Art
Stockwell Street, Leek, Staffordshire ST13 6DP

Lewes Technical College
Mountfield Road, Lewes, East Sussex BN7 2AH

Lincolnshire College of Agriculture and Horticulture
Caythorpe Court, Grantham, Lincolnshire NG32 3EP

Lindsay College of Agriculture
Riseholme, Lincoln LN2 2LG

Macclesfield College of Further Education
Park Lane, Macclesfield, Cheshire SK11 8LF

Merrist Wood Agricultural College
Worplesden, Nr Guildford, Surrey GU3 3PE

Mid-Cheshire College of Further Education
Hartford Campus, Northwich, Cheshire CW8 1LJ

Mid-Kent College of Further and Higher Education
Horsted, Maidstone Road, Chatham, Kent ME5 9UQ

Newbury College of Further Education
Oxford Road, Newbury, Berkshire RG13 1PQ

Norfolk College of Agriculture and Horticulture
Easton, Norwich, Norfolk NR9 5OX

Norfolk College of Art and Technology
Tennyson Avenue, Kings Lynn, Norfolk PE30 2QW

North Devon College
Sticklepath, Barnstaple, Devon EX31 2BQ

North Oxfordshire Technical College and School of Art
Broughton Road, Banbury, Oxfordshire OX16 9QA

North Tyneside College of Further Education
Embleton Avenue, Wallsend NE28 9NJ

Northamptonshire College of Agriculture
Moulton, Northampton NN3 1RR

Careers in Agriculture

Northumberland College of Agriculture
Ponteland, Newcastle upon Tyne NE20 0AQ

Norton-Radstock Technical College
South Hill Park, Radstock, Bath, Avon BA3 3RW

Nottinghamshire College of Agriculture
Brackenhurst, Southwell NG25 0QF

Oswestry College
College Road, Oswestry, Shropshire SY11 2SA

Pershore College of Agriculture
Avonbank, Pershore WR10 3JP

Pickering Agricultural Centre
Swainsea Lane, Pickering, North Yorkshire YO18 8NE

Plumpton Agricultural College
Plumpton, Lewes, East Sussex BN7 3AG

Rockingham College of Further Education
Wath upon Dearne, Rotherham, South Yorkshire S63 6PX

Royal Agricultural College
Cirencester, Gloucestershire

Rycotewood College
Priest End, Thame, Oxfordshire OX9 2BR

Seale Hayne College
Newton Abbot, Devon TQ12 6NQ

Shrewsbury College of Arts and Technology
London Road, Shrewsbury, Salop SY2 6PR

Shropshire Farm Institute
Walford, Baschurch, Shrewsbury SY4 2HL

Shuttleworth Agricultural College
Old Warden Park, Biggleswade, Bedfordshire SG18 9DX

Silsoe College
Room 28, Silsoe, Bedford MK45 4DT

Somerset College of Agriculture and Horticulture
Cannington, Nr Bridgwater, Somerset TA5 2LS

South Kent College of Technology
The Grange, Shornecliffe Road, Folkestone CT20 2NA

South Oxfordshire Technical College
Deanfield Avenue, Henley on Thames, Oxfordshire RG9 1UH

Staffordshire College of Agriculture
Rodbaston, Penkridge, Stafford ST19 5PG

Strode College
Church Road, Street, Somerset BA16 0AB

Useful Addresses

Wakefield District College
Margaret Street, Wakefield, West Yorkshire WF1 2DH

Walford College of Agriculture
Walford, Baschurch, Shrewsbury SY4 2HL

Warwickshire College of Agriculture
Moreton Hall, Moreton Morrell, Warwick CV34 9BL

West Kent College of Further Education
Brook Street, Tonbridge TN9 2PW

West Oxfordshire Technical College
Holloway Road, Witney, Oxfordshire OX8 7EE

West Sussex College of Agriculture and Horticulture
Brinsbury, North Heath, Pulborough, West Sussex RH20 1DL

Weston Super Mare Technical College and School of Art
Knightstone Road, Weston Super Mare, Avon BS23 2AL

Worcestershire College of Agriculture
Hindlip, Worcester WR3 8SS

Writtle Agricultural College
Writtle, Chelmsford, Essex CM1 3RR

Yeovil College
Ilchester Road, Yeovil, Somerset BA21 3BA

Wales

Brecon College of Further Education
Penlan, Brecon, Powys LD3 9SR

Carmarthen Technical and Agricultural College
Pibwrlwyd, Carmarthen, Dyfed SA31 2NH

Glynllifon College of Further Education
Clynnog Road, Caernarvon, Gwynedd LL54 5DU

Llysfasi College of Agriculture
Ruthin, Clwyd LL15 2LB

Coleg Meirionnydd
Dolgellau, Gwynedd LL40 2YF

Mid-Glamorgan College of Agriculture and Horticulture
Pencoed, Bridgend, Mid-Glamorgan CF35 5LG

Montgomery College of Further Education
Newton, Powys SY16 1BE

Pembrokeshire Technical College
College Campus, Haverfordwest SA61 1TG

Coleg Pencraig
Llangefni, Isle of Anglesey, Gwynedd LL77 7HY

Careers in Agriculture

The Usk College of Agriculture
Usk, Gwent NP5 1XL

Welsh Agricultural College
Joint Education Committee, Llanbadarn Fawr, Aberystwyth,
Dyfed SY23 3AL

The Welsh College of Horticulture
Northop, Nr Mold, Clwyd CH7 6AA

Scotland

Angus Technical College
Keptie Road, Arbroath, Angus DD11 3EA

Barony Agricultural College
Parkgate, Dumfries DG1 3NE

Borders Agricultural College
Newtown Street, Duns, Berwickshire TD11 3AE

Clinterty Agricultural College
Kinellar, Aberdeen AB5 0TN

East of Scotland College of Agriculture
Edinburgh School of Agriculture, West Mains Road,
Edinburgh EH9 3JG

Elmwood Agricultural and Technical College
Carslogie Road, Cupar, Fife KY15 4JB

Falkirk College of Technology
Grangemouth Road, Falkirk, Stirlingshire FK2 9AD

Galashiels College of Further Education
Melrose Road, Galashiels, Selkirkshire TD1 2AF

Henderson Technical College
Hawick, Roxburghshire TD9 7AW

Inverness Technical College
3 Longman Road, South Inverness IV1 1SA

Kilmarnock College
Holehouse Road, Kilmarnock, Ayrshire KA3 7AT

Kirkwall Further Education Centre
Grammar School, Kirkwall, Orkney KW15 1JG

Motherwell College
Dalzell Drive, Motherwell, Lanarkshire ML1 2DD

North of Scotland College of Agriculture
School of Agriculture, 581 King Street, Aberdeen AB9 1UD

Oatridge Agricultural College
Ecclesmachan, Broxburn, West Lothian EH52 6NQ

Useful Addresses

Perth College of Further Education
Brahan Estate, Crieff Road, Perth PH1 2NX

Reid Kerr College
Renfrew Road, Paisley, Renfrewshire PA3 4DR

Stranraer Agricultural Further Education Centre (Barony)
'Ryan', Lewis Street, Stranraer, Wigtownshire DF9 7LW

Thurso Technical College
Ormlie Road, Thurso, Caithness KW14 7EE

West of Scotland Agricultural College
Central Building, Auchincruive, Ayr KA6 5HW

Northern Ireland

Armagh Technical College
Lonsdale Street, Armagh BT61 7HN

Banbridge Technical College
Castlewellan Road, Banbridge, County Down BT32 4AY

Downpatrick Technical College
Downpatrick, County Down BT30 6ND

Dungannon Technical College
Circular Road, Dungannon, County Tyrone BT71 6BQ

Enniskillen Agricultural College
Irvinestown Road, Enniskillen, County Fermanagh BT74 6DN

Fermanagh College of Further Education
Fairview Avenue, Enniskillen, County Fermanagh BT74 6AE

Greenmount Agricultural and Horticultural College
Antrim BT41 4PU

Larne College of Further Education
32-34 Pound Street, Larne, County Antrim BT40 1SQ

Limavady Technical College
Limavady, County Londonderry BT49 0EX

Loughry College of Agriculture and Food Technology
Cookstown, County Tyrone BT80 9AA

The New Technical College
Coleraine Road, Ballymoney, County Antrim BT53 6BT

Omagh Technical College
Omagh, County Tyrone BT79 7AH

Portadown Technical College
Lurgan Road, Portadown, Craigaron, County Armagh BT63 5BL

Careers in Agriculture

Organisations Involved in Agriculture

This section includes agricultural associations, clubs, societies and other bodies involved in agriculture and able to supply information on the subject as well as advice in certain areas.

AFRC Food Research Institute
Colney Lane, Norwich NR4 7UA

Agricultural and Food Research Council
160 Great Portland Street, London W1N 6DT

Agricultural Training Board
32-34 Beckenham Road, Beckenham, Kent BR3 4PB

Agricultural Wages Board
Eagle House, Cannon Street, London EC4

The Biochemical Society
7 Warwick Court, High Holborn, London WC1R 5DP

British Sugar
PO Box 26, Oundle Road, Peterborough PE2 9QU

Business and Technician Education Council (BTEC)
Central House, Upper Woburn Place, London WC1H 0HE

Careers and Occupational Information Centre
5 St Andrews Place, London NW1 4LB

City and Guilds of London Institute
76 Portland Place, London W1N 4AA

Civil Service Commission
Alençon Link, Basingstoke, Hampshire RG21 1JB

Department of Agriculture and Fisheries for Scotland
Chesser House, Gorgie Road, Edinburgh EH11 3AW

Department of Education and Science
Information Division, Elizabeth House, York Road, London SE1 7PH

Department of Employment Training Services Agency
Ebury House, Ebury Bridge Road, London SW1

Agricultural Engineering Information
Engineering Careers Information Service
54 Clarendon Road, Watford WD1 1LB

Fish Farming Committee
National Farmers Union, Agriculture House, London SW1X 7NJ

Forestry Commission
231 Corstorphine Road, Edinburgh EH12 7AT

Useful Addresses

Institute of Agricultural Engineers
West End Road, Bedford MK45 4DU

Institute of Agricultural Secretaries
1 Main Street, Elloughton, Brough, North Humberside HU15 1JN

Institute of Animal Technicians
5 South Parade, Somerton, Oxford OX2 7JL

Institute of Brewing
33 Clarges Street, London W1Y 8EE

Institute of Chartered Secretaries and Administrators
16 Park Crescent, London W1N 4AH

Institute of Fisheries Management
109 Burgh Road, Carlisle, Cumbria

Institute of Food Science and Technology in the UK
4th Floor, 20 Queensberry Place, London SW7 2DR

Landscape Institute
12 Carlton House Terrace, London SW1Y 5AH

Ministry of Agriculture, Fisheries and Food
Whitehall Place (West Block), London SW1A 2HH
(The local office of the Ministry of Agriculture can be found in your local telephone directory.)

Manpower Services Commission
Selkirk House, 166 High Holborn, London WC1V 6PE

The National Association of Agricultural Contractors
Huts Corner, Tilford Road, Hindhead, Surrey GU26 6SF
(for contract work information)

National Farmers Union
The union has several addresses, each covering different farming regions. The areas relevant to each office are shown in brackets.

Information Officer, Agricultural House, Salters Lane South, Haughton-le-Skerne, Darlington, Co Durham DL1 2AA (Cumbria, Durham, Northumberland, Yorkshire – East and West Ridings)

Information Officer, Agriculture House, 119 Nantwich Road, Crewe, Cheshire CW2 6BD (Cheshire, Derbyshire, Lancashire, Yorkshire – West Riding)

Information Officer, 49-57 High Street, Droitwich, Worcestershire (Berkshire, Herefordshire, Leicestershire, Northamptonshire, Nottinghamshire, Oxfordshire, Rutland, Shropshire, Staffordshire, Warwickshire and Worcestershire)

Information Officer, Agriculture House, George Street, Huntingdon, Cambridgeshire (Bedfordshire, Buckinghamshire, Cambridgeshire, Essex, Hertfordshire, Lincolnshire, Norfolk and Suffolk)

Careers in Agriculture

Information Officer, Agriculture House, 31 Trull Road, Taunton, Somerset TA1 4QG (Cornwall, Devon, Dorset, Gloucestershire, Somerset and Wiltshire)

Information Officer, Agriculture House, Station Road, Liss, Hampshire GU33 7AR (Central Southern England, Hampshire, Isle of Wight and West Sussex)

Information Officer, Agriculture House, High Street, Cranbrook, Kent TN17 3EN (Kent and East Sussex)

Information Officer, Agriculture House, 19-21 Cathedral Road, Cardiff CF1 9LJ (Wales)

National Institute of Agricultural Botany
Huntingdon Road, Cambridge CB3 0LE

The Education Officer
National Joint Apprenticeship Council for the Agricultural Machinery Trade
Church Street, Rickmansworth, Hertfordshire (for career details and apprenticeships schemes)

National Proficiency Tests Council
YFC Centre, National Agricultural Centre, Kenilworth, Warwickshire CV8 2LZ

National Union of Agricultural and Allied Workers
Headland House, 308 Gray's Inn Road, London WC1X 8DS

Royal Agricultural Society of England
National Agricultural Centre, Kenilworth, Warwickshire CV8 2LZ

Royal Association of British Dairy Farmers
Robarts House, Rossmore Road, London NW1 6NP

The Soil Association
Walnut Tree Manor, Haughley, Stowmarket, Suffolk IP14 3RS

Enquiries Unit
Voluntary Service Overseas
9 Belgrave Square, London SW1X 8PW

Young Farmers' Clubs
YFC Centre, National Agricultural Centre, Kenilworth, Warwickshire CV8 2LZ

COLLINGHAM TUTORS,
23, COLLINGHAM GARDENS,
LONDON, S.W.5.